Introduction to
Noise-Resilient Computing

Synthesis Lectures on Digital Circuits and Systems

Editor
Mitchell A. Thornton, *Southern Methodist University*

The Synthesis Lectures on Digital Circuits and Systems series is comprised of 50- to 100-page books targeted for audience members with a wide-ranging background. The Lectures include topics that are of interest to students, professionals, and researchers in the area of design and analysis of digital circuits and systems. Each Lecture is self-contained and focuses on the background information required to understand the subject matter and practical case studies that illustrate applications. The format of a Lecture is structured such that each will be devoted to a specific topic in digital circuits and systems rather than a larger overview of several topics such as that found in a comprehensive handbook. The Lectures cover both well-established areas as well as newly developed or emerging material in digital circuits and systems design and analysis.

Introduction to Noise-Resilient Computing
S.N. Yanushkevich, S. Kasai, G. Tangim, A.H. Tran, T. Mohamed, and V.P. Shmerko
2013

Atmel AVR Microcontroller Primer: Programming and Interfacing, Second Edition
Steven F. Barrett and Daniel J. Pack
2012

Representation of Multiple-Valued Logic Functions
Radomir S. Stankovic, Jaakko T. Astola, and Claudio Moraga
2012

Arduino Microcontroller: Processing for Everyone! Second Edition
Steven F. Barrett
2012

Advanced Circuit Simulation Using Multisim Workbench
David Báez-López, Félix E. Guerrero-Castro, and Ofelia Delfina Cervantes-Villagómez
2012

Embedded Systems Design with the Atmel AVR Microcontroller: Part I
Steven F. Barrett
2009

Embedded Systems Interfacing for Engineers using the Freescale HCS08 Microcontroller II: Digital and Analog Hardware Interfacing
Douglas H. Summerville
2009

Designing Asynchronous Circuits using NULL Convention Logic (NCL)
Scott C. Smith and JiaDi
2009

Embedded Systems Interfacing for Engineers using the Freescale HCS08 Microcontroller I: Assembly Language Programming
Douglas H.Summerville
2009

Developing Embedded Software using DaVinci & OMAP Technology
B.I. (Raj) Pawate
2009

Mismatch and Noise in Modern IC Processes
Andrew Marshall
2009

Asynchronous Sequential Machine Design and Analysis: A Comprehensive Development of the Design and Analysis of Clock-Independent State Machines and Systems
Richard F. Tinder
2009

An Introduction to Logic Circuit Testing
Parag K. Lala
2008

Pragmatic Power
William J. Eccles
2008

Multiple Valued Logic: Concepts and Representations
D. Michael Miller and Mitchell A. Thornton
2007

High-Speed Digital System Design
Justin Davis
2006

Introduction to Logic Synthesis using Verilog HDL
Robert B.Reese and Mitchell A.Thornton
2006

Microcontrollers Fundamentals for Engineers and Scientists
Steven F. Barrett and Daniel J. Pack
2006

Introduction to Noise-Resilient Computing

S.N. Yanushkevich, S. Kasai, G. Tangim, A.H. Tran, T. Mohamed, and V.P. Shmerko

ISBN: 978-3-031-79854-2   paperback
ISBN: 978-3-031-79855-9   ebook

DOI 10.1007/978-3-031-79855-9

A Publication in the Springer series
SYNTHESIS LECTURES ON DIGITAL CIRCUITS AND SYSTEMS

Lecture #40
Series Editor: Mitchell A. Thornton, *Southern Methodist University*
Series ISSN
Synthesis Lectures on Digital Circuits and Systems
Print 1932-3166   Electronic 1932-3174

Introduction to
Noise-Resilient Computing

S.N. Yanushkevich
University of Calgary, Canada

S. Kasai
Hokkaido University, Japan

G. Tangim, A.H. Tran, T. Mohamed, and V.P. Shmerko
University of Calgary, Canada

SYNTHESIS LECTURES ON DIGITAL CIRCUITS AND SYSTEMS #40

ABSTRACT

Noise abatement is the key problem of small-scaled circuit design. New computational paradigms are needed; as these circuits shrink, they become very vulnerable to noise and soft errors. In this lecture, we present a probabilistic computation framework for improving the resiliency of logic gates and circuits under random conditions induced by voltage or current fluctuation. Among many probabilistic techniques for modeling such devices, only a few models satisfy the requirements of efficient hardware implementation; specifically, Boltzman machines and Markov Random Field (MRF) models. These models have similar built-in noise-immunity characteristics based on feedback mechanisms. In probabilistic models, the values 0 and 1 of logic functions are replaced by degrees of beliefs that these values occur. An appropriate metric for degree of belief is probability. We discuss various approaches for noise-resilient logic gate design, and propose a novel design taxonomy based on implementation of the MRF model by a new type of binary decision diagram (BDD), called a cyclic BDD. In this approach, logic gates and circuits are designed using 2-to-1 bi-directional switches. Such circuits are often modeled using Shannon expansions with the corresponding graph-based implementation, BDDs. Simulation experiments are reported to show the noise immunity of the proposed structures. Audiences who may benefit from this lecture include graduate students taking classes on advanced computing device design, and academic and industrial researchers.

KEYWORDS

nanotechnology, nanostructure, nanodevice, fluctuation, logic gate, noise-tolerance, fault-tolerance, Bayesian network, Markov random field, Hopfield model, Boltzmann machine, binary decision diagram

Contents

Preface

The authors are pleased to introduce all readers to the advanced topic of "noise-tolerant computation" or "noise- and fault-resilient computing."

TOWARD PREDICTABLE TECHNOLOGIES. This book is motivated by emerging gate-level solutions for implementing various noise- and fault-resilient computing platforms, created using contemporary and forthcoming nano- and molecular technologies. Insights from conventional design as well as probabilistic techniques are both valuable sources for the further development of this new generation of computing devices. In this book, multi-feedback Programmable Logic Arrays (PLAs) and cyclic binary decision diagrams (BDDs) are adopted for the implementation of probabilistic models of logic gates, and, thus, for prototyping of future computing nano-devices and nano-memory. The noise- and fault-resilient computing paradigm plays an essential role in the development of such novel computing structures based on technological advances in deep sub-micron technology.

AUDIENCE. This book presents the necessary computing concepts, techniques, and tools in a form ideally suitable to graduate students and developers specializing in the design of new computing devices. The organization of the book offers a great deal of flexibility for use in graduate courses related to advanced design. This book is based on selected lectures, adopted for graduate courses in electrical and computer engineering programs at the University of Calgary (Canada) and Hokkaido University (Japan). Standard prerequisites for these programs include courses on fundamentals of logic design, based on the textbooks that include contemporary data structures such as binary decision diagram (in particular, [153]). With the advent of new technologies, a major shift has occurred in device reliability studies, motivating the project of how to provide for reliable computation on unreliable logic devices. This trend has produced a need for basic knowledge of (a) Computer science and computing engineering, (b) Engineering statistics, (c) Electrical and logic circuit design, and (d) Electronics. This book is motivated by interdisciplinary prospects for the development of a new generation of computer devices and systems.

STRUCTURE OF THE BOOK. This book contributes to hardware design of noise-tolerant integrated devices. The book consists of six chapters, organized as follows:

Chapter 1 *Introduction to probabilistic computation models.* The focus of this chapter is on the hardware implementation of noise-tolerant models.

Chapter 2 *Nano-scale circuits and fluctuation problems.* Noise and variation in contemporary nanoelectronic devices is the subject of this chapter.

Chapter 3 *Estimators and metrics.* Specific metrics, required for the design and performance evaluation of probabilistic devices, are described in this chapter.

Chapter 4 *MRF-based models of logic gates.* The chapter reviews the PLA-based implementation of the Markov Random Field (MRF) approach, and focuses on cyclic BDD implementation. The latter is ideal for technologies offering simple bi-directional 2-to-1 switches. This chapter is an extension of the results initially introduced in [100].

Chapter 5 *Neuromorphic models.* The results of the discrete Hopfield model of logic gates and its extension—the Boltzmann machine—are introduced.

Chapter 6 *Noise-tolerance via error correcting.* The fault-tolerant design of logic gates using error correcting coding (ECC) and BDD-based implementation are emphasized in this chapter. This chapter should be considered as implementation aspects (using a single-electron quantum logic circuit based on Schottky wrap gate control of a GaAs nanowire) of an earlier study [8].

Most of the models, introduced in Chapters 3-6, have been simulated using the SPICE package with the 16-nm Berkeley CMOS technology (http://ptm.asu.edu/).

MAJOR CONTRIBUTION AND NEW HORIZONS. This book emphasizes the basic principles of computing under uncertainty. The following key features distinguish this book from other known approaches to noise- and fault-tolerant design:

(a) Iterative design based on feedback computing paradigms, such as MRF and Hopfield models, provides for a reasonable noise-resilient hardware implementation of elementary logic gates.

(b) Analog designs are preferable for the implementation of noise-resilient computing.

(c) Noise-resilient designs are acceptable for high-performance extension of classical computing paradigms toward 3D computing structures [125, 154].

HOW THIS BOOK WAS CREATED. This book was created using selected lectures for graduate students at the University of Calgary, Department of Electrical and Computer Engineering, Canada, the Graduate School of Information Science and Technology, and Research Center for Integrated Quantum Electronics, Hokkaido University, Japan. Dr. Yanushkevich visited Hokkaido University in December 2011. This visit was supported by NSERC (Canada) and the Japanese Society for the Promotion of Science (JSPS) and hosted by Dr. Seiya Kasai. His group was the first to design and fabricate single electron quantum logic circuits based on Schottky wrap gate control of a GaAs nanowire. These results were reported, in particular, in the seminal paper by Dr. S. Kasai and Dr. H. Hasegawa, "A single electron BDD quantum logic circuit based on Schottky wrap gate control of a GaAs nanowire hexagon," published in IEEE Electron Device Letters more than ten years ago (vol. 23, no. 8, 2002), and later publication [38].

Dr. Yanushkevich's visit to RCIQE in 2011 served as a catalyst for further collaboration on addressing this problem. The MRF models presented in this book are candidates for implementing using GaAs nanowire devices at the RCIQE at Hokkaido University.

Dr. Seiya Kasai works at the Research Center for Integrated Quantum Electronics at Hokkaido University. The group has experimentally proven that binary decision diagrams (BDDs) are a feasible data structure for computing at the nanoscale. The authors also emphasized the probabilistic nature of computation.

Through the JSPS Fellowship Program, Dr. Yanushkevich also visited the Department of Computer Science and Electronics at the Kyushu Institute of Technology, Iizuka.

Dr. Tsutomu Sasao, IEEE Fellow, is a Professor in the Department of Computer Science and Electronics at the Kyushu Institute of Technology. He is an international leader in the logic design of discrete devices. Discussions with him and his graduate students invoked some fruitful ideas on generalizing probabilistic models toward the multi-valued logic (MVL). His research areas include logic design and switching theory, representations of logic functions, and MVL. Dr. Sasao is a long-term member of the MVL community, who has served as the Symposium and Program Chair for the Annual IEEE Int. Symposium on MVL many times.

Dr. Sasao's textbook "Switching Theory for Logic Synthesis," published by Kluwer in 1999, is a classic text that is used worldwide. Dr. Sasao received the NIWA Memorial Award in 1979, and multiple Distinctive Contribution Awards from the IEEE Computer Society MVL-TC. In particular, the problem of planarization of decision diagrams, discussed in this book, was addressed by Dr. Sasao for MVL back in 1995 (T. Sasao and J. T. Butler, "Planar Multiple-Valued Decision Diagrams," IEEE Int. Symposium on MVL, Bloomington, Indiana, 1995; it was selected as a Distinctive Contribution Award, and was published as an invited paper in *Journal on MVL* (vol. 1, no. 1, 1996). It also inspired our research group to later develop linear decision diagrams ("Linearity of word-level models: New understanding," *Proc. IEEE/ACM 11 Int. Workshop on Logic and Synthesis* by S. N. Yanushkevich, et al., New Orleans, LA, 2002.) Planar approaches have been discussed in both groups with respect to implementation on quantum logic circuits.

Dr. S. N. Yanushkevich visited the Research Center for Integrated Quantum Electronics at Hokkaido University.

On behalf of the authors of this collection of lectures
Svetlana N. Yanushkevich
Calgary, Canada – Sapporo – Iizuka, Japan
December 2011–December 2012

Acknowledgments

This work was partially supported by the Natural Sciences and Engineering Research Council (NSERC) of Canada, and the Japanese Society for the Promotion of Science (JSPS). Dr. S. Yanushkevich and Dr. S. Kasai acknowledge the JSPS Invitation Fellowship program. Mr. G. Tangim acknowledges the Information and Communication Technologies Recruitment Scholarship, University of Calgary. Mr. A. H. Tran acknowledges the Natural Sciences and Engineering Research Council of Canada. Mr. T. Mohamed acknowledges the iCore program (Alberta, Canada).

Also we would like to acknowledge the editor of this lecture series Dr. Mitchell A. Thornton.

S.N. Yanushkevich, S. Kasai, G. Tangim, A.H. Tran, T. Mohamed, and V.P. Shmerko
January 2013

CHAPTER 1

Introduction to probabilistic computation models

Murphy's principle of probabilistic computation

You can't be 100 percent right, but you can be 100 percent wrong

In this chapter, we present a brief introduction to the area of logic design known as *probabilistic computing*, or *computing under uncertainty*, the uncertainty being induced by various types of variability. These "umbrella" terms refer to various aspects of probabilistic data representation and analysis, random data generation and approximation, modeling and computing, using statistical estimations. For example, what is common to all *error-resiliency*, *error-tolerant*, and *noise-tolerant* techniques is that they permit the gates or circuits to make errors, and correct them so that the circuit level specifications are relaxed, thereby saving energy. These designs are characterized by performance metrics of a statistical nature, such as signal-to-noise ratio (SNR), probability of error detection, and bit error-rate (BER). These techniques imply that the gates and circuits need not be 100% correct as long as the application requirements are met.

The focus of our interest is on probabilistic models for logic gates, which are the simplest devices for computing elementary logic functions, such as NOT, AND, OR, NAND, NOR, and EXOR. These models must be acceptable for simple hardware implementation. Among the models proposed so far, only the MRF model and Hopfield model have been simulated and investigated to an extent that permits them to be considered as candidates for noise-tolerant logic gates design.

1.1   WHY DO WE NEED PROBABILISTIC MODELS?

There are various fields in which probabilistic computing is a key technique. For example, in communication, the best error correcting technique, known as turbo coding, is based on probabilistic data processing [11]. The devices and systems, known as *belief networks*, are widely used in decision-making, medical diagnosis, image and voice processing, robotics, and control systems [52]. In forthcoming technologies—in particular, molecular electronics—probabilistic computing platforms are needed to deal correctly with the stochastic nature of molecular device behavior.

1.1.1   NOISE

There are two main sources of noise: (a) variations in parameters and (b) scaling of the supply voltage. In deep submicron and nanoscale devices, various phenomena, such as capacitive electrostatics, thermal fluctuations, and magnetic interference, cause intrinsic noise.

Variability of circuit parameters has become a key factor in mainstream digital design. Variations can be classified as systematic (having a quantitative relationship with a source), random (unpredictable), drift-related (due to aging and temperature variation), and jitter-related (from voltage variations or crosstalk). In this chapter, we will focus on random variations. These variations affect transistor ON and OFF currents, which influence circuit delay behavior and energy dissipation characteristics. These variations can cause a circuit to malfunction, especially at low voltages. Variations have traditionally been handled by using margining. However, the ever-increasing margins severely degrade the performance and power of circuit. Therefore, variation-tolerant designs are becoming more important; in particular, noise-tolerant and fault-tolerant designs.

Many devices and systems are required to operate at low power, in order to extend battery life, such as portable medical processors and implants, distributed sensor networks, and active radio-frequency identification devices [114]. One commonly employed technique in the design of such devices and systems is aggressive voltage scaling. But while scaling the supply voltage helps to reduce internal noise (such as crosstalk noise), it makes the device more sensitive to external (not scaled) noise sources. Reducing the supply voltage of a CMOS device improves its energy dissipation characteristics, but the device also becomes more sensitive to variations in its parameters, such as the transistor threshold [5]. The threshold voltage is determined by the number and location of dopant atoms, implanted in the channel or halo region. The ion implantation is a stochastic process, leading to random dopant fluctuations. For example, the standard deviation in threshold voltage for a minimum-sized device is approximately 10 mV in a 90 nm process, 30 mV in a 50 nm process, and 40 mV in a 25 nm process [3]. As suggested in the literature [64], there must also be a design trade-offs between energy, performance, and probability of correct computing.

New computing paradigms are needed to deal with the effects of scaling the supply voltage. As an example of such a new paradigm, consider the technique of a *resilient power management* under sources of uncertainty [54]. This technique is based on stochastic processing and models, adopted from various fields; in particular, from pattern classification and intelligent system design [27, 40]. The key concept in estimation and decision making is *belief*, which is represented by a probability distribution over all possible values of parameters (variables). In Table 1.1, a conventional logic gate and its probabilistic model are compared in terms of the design process; in particular, performance, formal representation, optimization, propagation mechanism, and metrics.

In 1952, von Neumann wrote: "...Error should be treated by thermodynamical methods, and be the subject of a thermodynamical theory, as information has been ..." Von Neumann's goal was to subject noisy computation to the same thermodynamical treatment that communication had received known as information (Table 1.1, see "Metric"). Specifically, the information carried by a signal decays when the signal is corrupted by noise. In such a model, a message is transmitted over

Table 1.1: Comparison of characteristics and metrics for performance evaluation of a conventional logic gate, and its design based on probabilistic model.

Characteristic	Conventional design	Probabilistic based model
Performance	Power consumption, area, delay	Number of iterations to converge
Formalization	Logic function	Joint probability distribution (global description), conditional distribution and marginal distributions (local description)
Optimization	Decomposition of logic function	Factorization of joint probability distribution
Propagation mechanism	Binary signal propagation	Probability propagation
Metric	Number of elements, depth of circuit, variation of parameters	Signal-to-noise ratio (SNR), bit-error rate (BER), Kullback-Leibler divergence (KLB), entropy, correlation, mutual information

a noisy channel, as well as when a *noisy logic gate* performs computation. We say that a gate fails if its output is incorrect (produces 1 instead of 0 or vice versa). A noisy logic gate fails with probability bounded by $p \in (0, 1/2)$ and gate is reliable with probability $1 - p$.

1.1.2 IDEAL NOISE-FREE CONDITIONS

Deterministic description of logic gates is reasonable when they operate on noise-free signals, that is, when noise can be neglected. An example of a noise-free environment is given in Fig. 1.1(a) where the AND gate operates with binary signals, $x_1, x_2 \in \{0, 1\}$, and produces the binary signal, $f = x_1 \wedge x_2$, $f \in \{0, 1\}$, accordingly to its truth table.

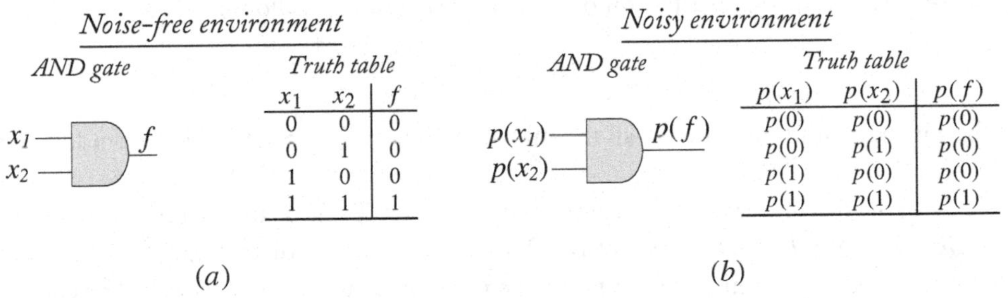

Noise-free environment

AND gate   *Truth table*

x_1	x_2	f
0	0	0
0	1	0
1	0	0
1	1	1

Noisy environment

AND gate   *Truth table*

$p(x_1)$	$p(x_2)$	$p(f)$
$p(0)$	$p(0)$	$p(0)$
$p(0)$	$p(1)$	$p(0)$
$p(1)$	$p(0)$	$p(0)$
$p(1)$	$p(1)$	$p(1)$

(a)   (b)

Figure 1.1: The logic AND gate description of a noise-free (a) and noisy (b) environment.

In logic design, uncertainty is defined by unspecified values of logic functions. Specifically, there are combinations that never appear in the inputs, or in the cases where the value of the function can be either 0 or 1. Combinations of inputs, for which the value of a function is not specified, are called *don't care* combinations. A function with don't cares is an incompletely specified function; that is, such a logic function contains uncertainty. Logic circuit design relies on the use of don't cares as degrees of freedom for optimization.

1.1.3 NOISY OPERATION CONDITIONS

The focus of our study is the behavior of logic gates with noisy input signals. Specifically, when noise is allowed, input signals, $x_1, x_2 \in \{0, 1\}$, are applied to a logic gate with a certain level of probability, $p(x_1)$ and $p(x_2)$ (Fig. 1.1b). Probability is used here to quantify the likelihood that a measurement falls within some range of values. Probabilistic models assume that correct output signals are calculated with a certain level of probability. In the case of the AND gate, $p(f) = p(x_1)p(x_2)$. For example, "0" and "1" can be understood as random variations of voltage or current around the values, corresponding to "0" and "1."

The probability of an event, such as a logical "0" or "1" at the output of gate, can be conditioned on knowing that another event has taken place at the inputs of gate. Conditional probabilities incorporate information about the occurrence of another event. The behavior of a logic gate in a probabilistic environment is completely described by *probability distribution functions*. Input probabilities, $p(x_1)$ and $p(x_2)$, are derived from these functions in the form of a *probabilistic truth table*. For simplicity, it is also often assumed that the input signals are uncorrelated and independent.

However, it is difficult to describe a gate using this approach, because the probability distributions are unknown, or their formal description is very complicated and intractable in practice. Unknown distributions are often approximated by uniform distributions of uncorrelated signals. This is a very rough probabilistic approximation. In this book, we consider probabilistic models of logic gates based on the *joint probability distribution function*, which is a function of the random inputs and outputs of a logic gate.

Recall that the probability distribution of a random variable X is a description of the set of the probabilities associated with the possible values for X. If X and Y are two discrete random variables, the joint probabilistic distribution of X and Y is defined as follows:

$$f(x, y) = P(X = x, Y = y).$$

The joint distribution contains all the information possible about the joint random variables X and Y.

In the discrete case, the marginal distributions of X and Y are calculated as $g(x) = \sum_y f(x, y)$ and $h(y) = \sum_x f(x, y)$, respectively. That is, the marginal distributions of X, $g(x)$ and Y, $h(y)$, are calculated by summing $f(x, y)$ over the values of Y and X, respectively. For continuous random variables, the marginal distribution of X and Y are calculated as $g(x) = \int_{-\infty}^{\infty} f(x, y)dy$ and $h(y) = \int_{-\infty}^{\infty} f(x, y)dx$, respectively, (Fig. 1.2a). The *conditional distribution* of the random variable Y and X, given that $X = x$ and $Y = y$, are $f(y|x) = \frac{f(x,y)}{g(x)}$ for $g(x) > 0$ and $f(x|y) = \frac{f(x,y)}{h(y)}$ for $h(y) > 0$, respectively. The random variables X and Y are *statistically independent* if and only if their joint distribution, $f(x, y)$, can be represented by the product of their marginal distributions, $f(x, y) = g(x)h(y)$. Generalization for three or more dimensions is straightforward (Fig. 1.2b). For more details, see [84].

For example, for a two-input AND gate (Fig. 1.1b), the joint distribution function is defined as $g(x_1, x_2, f)$. In a noisy environment, this is a three-dimensional function, which is unaccept-

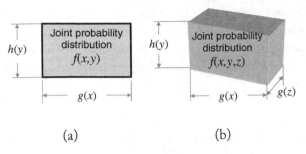

(a) (b)

Figure 1.2: (a) A two-dimensional probability distribution $f(x, y)$ and its marginal distributions $g(x)$ and $h(y)$ of random variables X and Y, respectively; (b) a three-dimensional distribution $f(x, y, z)$

able for computation because of its high complexity. However, appropriate models are known for approximation and simplified computing.

A similar technique for probabilistic analysis using a graphical interpretation is introduced in Fig. 1.3. If the probability of observing a logic "0" at the input of the binary inverter as $p(x = 0) = 0.2$, then the probability of a logic "1" is $p(x = 1) = 1 - 0.2 = 0.8$. Therefore, at the output of the inverter we observe the reverse: $p(y = 0) = 0.8$ and $p(y = 1) = 0.2$ (Fig. 1.3a).

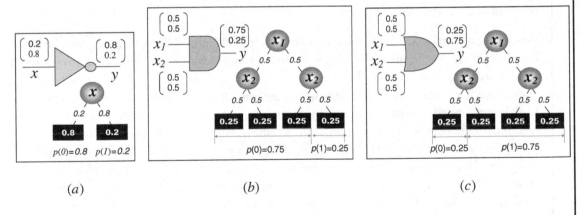

(a) (b) (c)

Figure 1.3: Probabilistic analysis using decision trees for logic gates: (a) NOT gate, (b) AND gate, and (c) OR gate.

1.2 PROBABILISTIC TECHNIQUES AND MODELS

The term "probabilistic computing" has multiple meanings [11, 28, 50, 72, 88, 90, 137]. Different methodologies are used in designing devices and systems for probabilistic computing. Some probabilistic techniques have been well known in logic design for a long time, such as the Gaines noise-tolerant adder and multiplier [34], Monte-Carlo simulation [66], and the entropy criterion for optimization of logic networks [2, 141]. In some cases, designers have a need for more detailed modeling of the probabilistic behavior of a computing device. Most such models employ the *principle of local computation*, and iteratively compute design parameters, accumulating statistics in order to make a final decision.

In a seminal paper [143], von Neumann introduced the model of *noisy logic circuit* composed of *noisy logic gates*, that produce independently with some probability a 0 instead of 1, or vice versa, in an attempt to capture the limitations of such computation. The classic problem formulation is as follows: Whether noisy circuit can compute the specified logic function as noisy-free circuit; and if so, at what cost in depth (latency)? Von Neumann provided the following solution: Every logic circuit with noiseless gates can be simulated by a circuit with noisy gates, whose depth is at most a constant times the depth of the noise-free circuit. Advantages and discussion of this model are still discussed [37, 112].

1.2.1 POPULAR PROBABILISTIC TECHNIQUES

The three most used techniques involving probabilistic models are:

(*a*) Stochastic operations,

(*b*) Information-theoretical models, and

(*c*) Probabilistic decision diagrams.

Stochastic operations.   In circuit design, the most popular probabilistic technique is called *stochastic computing* [15, 34, 50, 90, 137]. This is an inherently stochastic paradigm that always results in approximate solutions. This technique is defined as a method for designing low-precision digital devices, such as adders and multipliers, which are intended to be operated in the presence of noise. Data is represented by binary stochastic pulse streams, and statistical averaging is used to calculate the required function. In this computational scheme, the cell implements one-bit addition/subtraction or multiplication with some level of correctness. The cells can be grouped into clusters to operate with many bits. In stochastic encoding, a bit string value is carried by a stream of random impulses, and even significant losses of information ("pieces" of random pulses) are not critical, because information (logical 1s and 0s) is distributed in hundreds or thousands of pulses with random intervals between them.

Consider the two-input logic AND cell in arithmetic (integer-valued) form, $f = x_1 \wedge x_2 = x_1 \times x_2$. The carriers of x_1 and x_2 are deterministic pulses, and this fact allows the replacement of the switching variables x_1 and x_2 by the *independent* random variables X_1 and X_2, respectively. Let random pulse streams be the carriers of these random variables. Randomness is represented by the intervals between impulses of the same amplitude, equal to a logical value of 1. The output of the AND cell is the multiplication of these pulse streams. To evaluate the output, let us calculate the mean of the input streams; that is, $E[X_1]$ and $E[X_2]$. Because an AND gate implements the multiplication of independent random variables, the output is $E[X_1 \times X_2] = E[X_1] \times E[X_2]$.

Hence, the following similarity can be observed after replacing the deterministic pulses x_1 and x_2 in the AND cell by independent random pulse streams X_1 and X_2 with mean of $E[X_1]$ and $E[X_2]$, respectively: $x_1 \wedge x_2 = x_1 \times x_2 \Leftrightarrow E[X_1] \times E[X_2]$.

Let us define the computational model for the case when stochastic streams are applied to the inputs of the OR gate. The output of the OR gate is described by the arithmetic expression $f = x_1 \vee x_2 = x_1 + x_2 - x_1 \times x_2$. If the probabilities of pulses in random streams are equal, the similarity between the processing of deterministic and random signals using an OR function is expressed as

$$\overbrace{x_1 \vee x_2 = x_1 + x_2 - x_1 x_2}^{OR \text{ gate}} \Leftrightarrow \overbrace{E[X_1] + E[X_2] - \underbrace{E[X_1]E[X_2]}_{Unwanted\ term}}^{Model}$$

In order to consider this result as a stochastic addition, the unwanted value $E[X_1] \times E[X_2]$ must be suppressed, or its influence on the sum $E[X_1] + E[X_2]$ must be mitigated.

The precision of computing depends on the size of the stochastic sequence. In practice, the size varies from hundreds to thousands, depending on the required precision of computations. This effect can be evaluated by standard statistical techniques.

Information-theoretical model. The keystone of information theory is Shannon entropy. The uncertainty, eliminated by the measurement, can be represented in terms of Shannon entropy. In a digital system, information is measured in units called bits (binary digits). Entropy measures the average number of binary digits that are needed to code a given message with a set of symbols, each with a given probability. A computing system can be seen as a process of communication between computer components. The classical concept of information advocated by Shannon is the basis for this. Specifically, if the probability distribution of a random variable X is known, the Shannon entropy is defined as $H(X) = -\sum_{i=1}^{n} p(x_i) \log_2 p(x_i)$, where $-\log_2 p(x_i)$ is the amount of information gained after observing the event $X = x_i$ with probability $p(x_i)$. Shannon entropy is used as a criterion function for comparing uncertainty reduction techniques [2, 19, 20, 79, 141].

Probabilistic decision diagrams. Decision diagrams are efficient graphical data structures for computing logic functions, and can be used for probabilistic analysis of the latter. Let a switching function f be represented by a binary decision diagram. The node in this diagram (assigned with the i-th variable, x_i, and with outgoing edges corresponding to $f_{x_i=0}$ and $f_{x_i=1}$) computes the Shannon expansion $f = \overline{x}_i f_0 \vee x_i f_1$, where $f_0 = f_{x_i=0}$ and $f_1 = f_{x_i=1}$. This equation can be interpreted as a probability propagation:

$$p(f) = p_{x_i=0} p_{f_{x_i=0}} + p_{x_i=1} p_{f_{x_i=1}}$$

assuming that co-factors $\overline{x}_i$ and f_0 and x_i and f_1 are independent. In this case, the decision diagram becomes a *probabilistic decision diagram* [123, 133, 134, 154]. These have various applications, such as estimation of switching activity in low-power design [79].

1.2.2 PROBABILISTIC MODELS BASED ON LOCAL COMPUTATION

Probabilistic models are the most sophisticated subset of probabilistic techniques. The simplest form of probabilistic modeling is called *Monte-Carlo simulation* [66]. In practice, Monte-Carlo simulation is commonly used to assess the impact of parameter variation in circuits. However, this method requires considerable computational resources, and is inefficient for large circuits. Currently used various probabilistic models and techniques for logic network modeling include:

- *Bayesian belief propagation models* in which input and output signals are probabilities (real numbers) of the corresponding signals [52, 104],
- *Markov random field (MRF) models* [35, 100, 145] which are similar to Bayesian networks, but can describe cyclic dependencies, which Bayesian networks cannot, and
- *Neuromorphic models* [40, 75, 135], which are implemented by a typical Hopfield associative memory, utilizing the Boltzmann updating algorithm.

The common feature of these models is that they exploit knowledge of signal statistics, statistical signal processing techniques, and probabilistic measures of uncertainty. However, these models are different with respect to input and output data, algorithms, implementation, and applications. In particular, Bayesian models require causal representations of data. In contrast, the Markov random field model is based on non-causal data description. The Hopfield model (implemented as neural network of a particular configuration) can be converted to a Markov random field model (implemented as an operational unit), and vice versa [35, 40, 91].

These models exploit different built-in capabilities to adapt the parameters of models to changes in the surrounding environment and various undesirable phenomena, such as noise, non-uniformity, delays, interference, process variations, leakage, and soft errors due to particle hits.

1.2.3 NEAREST NEIGHBOR METHODOLOGIES

Nearest neighbor-based models employ local data representation and computing [27, 35, 42, 44]. As a rule, nearest neighbor methodologies utilize a graphical data structure. For example, nearest neighbor computing is a typical technique for image processing and pattern recognition problems [27]. The crucial step in the nearest neighbor approach is to calculate the proximity of the object of interest to each of the neighbor-objects. The measure of proximity may be, for example, distance, or may be a particular feature, such as cliques or complete subgraphs of causal (e.g., Bayesian networks) or non-causal (e.g., MRF and Boltzmann machine) graphical data structures.

Construction of a probabilistic model. There are several steps in the construction of a probabilistic model:

(a) Problem description using graphical structures, such as directed or undirected graphs (corresponding to causal or non-causal representations, respectively);

(b) Application of a method for proximity calculation (e.g., cliques, distance measures).

(*c*) Probabilistic description, such as a factored representation of a joint probability distribution; and

(*d*) Computing using maximum likelihood principle.

Embedding a logic function into a model.   The properties of a logic function are represented using (a) a graphical data structure, and (b) a mechanism for incorporation into the intermediate function, called the energy function. The graph configuration conveys the structural properties of the logic function via relations between inputs and outputs. For example, a two-input logic gate can be represented by a three-node complete graph. The rest of the properties of the logic function are represented by an intermediate function called the energy function. In a Bayesian network, for example, the causal graphical data structure is a copy of the logic network configuration. In a MRF or Boltzmann machine, a logic function is embedded using local graphical structures and the energy function.

1.2.4   BAYESIAN BELIEF PROPAGATION MODEL

Recalling Bayes's rule.   Let us think of event B_r as being a possible **hypothesis** about some subject matter (Fig. 1.4) [126]. Bayes's formula may be interpreted as follows: How opinions about these hypotheses held before the experiment, $P(B_r)$, should be modified by the evidence of the experiment. In Bayes's formula, $P(B_r|A)$ is called **revised**, or **posterior** probability. This rule is a simplified version "Bayes's theorem," a rule for updating previously held beliefs in the light of evidence. A previously held degree of belief in a hypothesis, called a **prior** probability $P(B_r)$, should be updated based on evidence A in order to arrive at a revised, or posterior probability $P(B_r|A)$. Note that in most cases belief in a statement is quantized into the states *true* and *false*.

Belief updating.   Bayes's theorem extends the rules of conditional probability and total probability and provides a method of dealing with inference and belief updating in uncertainty situations. To update a belief in a hypothesis, we need: (a) some belief in the hypothesis based on previous evidence, and (b) some new evidence that supports the hypothesis. The goal of belief updating is to determine the current belief in the hypothesis given new evidence and some previous belief. In probabilistic computation using belief updating, the evidence is distinguished as follows:

(*a*) *Positive evidence* which initiates a *reward* event; this event increases the total belief in a hypothesis, and

(*b*) *Negative evidence* which initiates a *penalty* event; this event decreases the total belief in a hypothesis.

Causal knowledge and modeling.   In the belief propagation model, any phenomenon must first be described in causal form, and then, using probabilistic relationships, transformed into a belief propagation network:

Bayes's rule

Assume a collection of mutually exclusive and exhaustive events $B_1, B_2, \ldots, B_k$, such that all of these events have a probability greater than zero of occurring, and one of events, B_i, $i = 1, 2, \ldots, k$, must occur. Then, for any other event A in S for which $P(A) > 0$, posterior probability

$$P(B_i|A) = \frac{P(B_i)P(A|B_i)}{\underbrace{\sum_{i=1}^{k} P(B_i)P(A|B_i)}_{\text{Total probability, } P(A)}}$$

Figure 1.4: Recalling Bayes' rule.

$$\text{Phenomenon} \quad \longrightarrow \quad \underbrace{\text{Causal model} \quad \overset{Design}{\longrightarrow} \quad \text{Belief network}}_{\text{Computing}}$$
$$\underbrace{\phantom{\text{Phenomenon} \longrightarrow \text{Causal model}}}_{Propositions}$$

Causal modeling attempts to resolve the question about possible causes so as to provide an explanation of phenomena (effects) as the result of previous phenomena (causes). Causal knowledge is modeled using causal networks, in which the nodes represent propositions (or variables), the arcs signify direct dependencies between the linked propositions, and the strengths of these dependencies are quantified by conditional probabilities. A Bayesian network is a type of belief network that captures the way the propositions relate to each other probabilistically.

Bayesian propagation model.   The simplest form of the belief propagation model is explained below. If k events $B_1, B_2, \ldots, B_k$ constitute a partition of the sample space S, such that $P(B_i) \neq 0$ for $i = 1, 2, \ldots, k$, then, for any events B_r and A of S such that $P(A) \neq 0$,

$$\underbrace{P(B_r|A)}_{\text{Posterior}} = \underbrace{P(A|B_r)}_{\text{Likelihood}} \times \frac{\overbrace{P(B_r)}^{\text{Prior}}}{\underbrace{P(A)}_{\text{Evidence}}}$$

where $r = 1, 2, \ldots k$; $P(B_r|A)$ is a *revised*, or *posterior*, probability; $P(A|B_r)$ is the *likelihood* of B_r with respect to A; $P(A)$ is the *evidence factor* (viewed as merely a normalization factor, guaranteeing that the posterior probabilities sum to one, as all good probabilities must). This belief propagation form, or Bayesian principle, determines how to update probabilities, once such a conditional probability structure has been adopted, given appropriate prior probabilities.

Factorization.   Let the nodes of a graph represent random variables $X = \{x_1, \ldots, x_m\}$, and the links between the nodes represent direct causal dependencies. A *Bayesian belief network* is based on a *factored* representation of a joint probability distribution,

$$
P(x_1, x_2, \ldots, x_n) = \overbrace{P(x_1) \times P(x_2|x_1) \times \ldots, P(x_n|x_1, \ldots, x_{n-1})}^{\text{Factored form}}
$$

$$
= \underbrace{\prod_i P(x_i|x_1, x_2, \ldots, x_{i-1})}_{\text{Chain rule}} \Leftrightarrow \underbrace{\prod_{i=1}^{m} P(x_i|\text{Par}(X_i))}_{\text{Graphical representation}} \qquad (1.1)
$$

where $\text{Par}(X_i)$ denotes a set of parent nodes of a random variable x_i. The nodes outside $\text{Par}(X_i)$ are conditionally independent of x_i. Reasoning with Bayesian networks is done by updating beliefs; that is, by computing the posterior probability distributions, given new information (*evidence*). The basic idea is that new evidence has to be propagated to the other parts of the network.

Bayesian belief networks for logic gates   In Table 1.2, the design cycle of Bayesian models for the AND, OR, and NOT logic gates is introduced. In the first step, a graphical model in the form of a directed graph is created (belief net I). Interpretation of this model in terms of conditional probability distribution results belief net II. In this model, the conditional probability table is assigned to each node using Equation 1.1, such as $p(y) = p(y|x_1)p(y|x_2)$. For example, given the input probabilities $p(x_1 = 0) = 0.2$ and $p(x_2 = 1) = 0.8$, the output probability is $p(y|x_1 = 0, x_2 = 1) = p(y|x_1 = 0)p(y|x_2 = 1) = 0.2 \times 0.8 = 0.16$.

An arbitrary causal network can be transformed into a *belief tree* (the last column in Table 1.2). From a computational point of view, these representations of Bayesian models are characterized by a need for only low-precision arithmetic operations and memory requirements for storing conditional probability tables.

Bayesian model for simple circuit.   A Bayesian model design taxonomy for two-input logic gates is presented in Table 1.2. Consider, for example, two connected OR and AND gates as shown in Fig. 1.5. Let us assume that preliminary statistical analysis tells us that the input probabilities of logical "0" and "1" are $p(x_1 = x_2 = x_3 = 0) = 0.2$ and $p(x_1 = x_2 = x_3 = 1) = 0.8$, respectively. We are interested in the output probabilities $p(y_2 = 0)$ and $p(y_2 = 1)$. A Bayesian model of this network is given in Fig. 1.6.

1.2.5   MARKOV RANDOM FIELD MODEL

The restoration of degraded images (allowing for noise) using the MRF model was reported in [35]. This model was explored for implementing the logic function in [10, 100]. The MRF, unlike the Hopfield model, does not require time-redundancy. It considers inputs and outputs of circuits as

Table 1.2: Design of Bayesian networks for the AND, OR, and NOT logic gates.

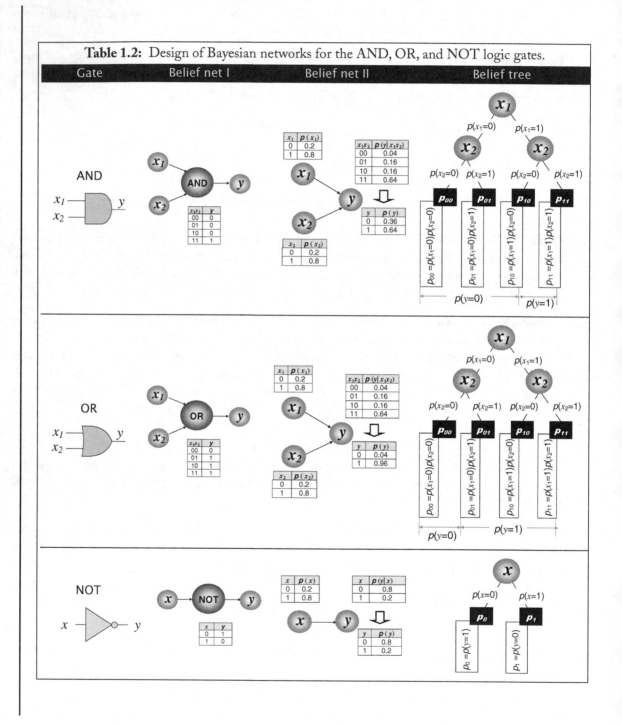

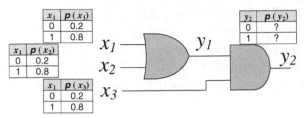

Figure 1.5: Probabilistic specification of a three-input single-output OR-AND logic network.

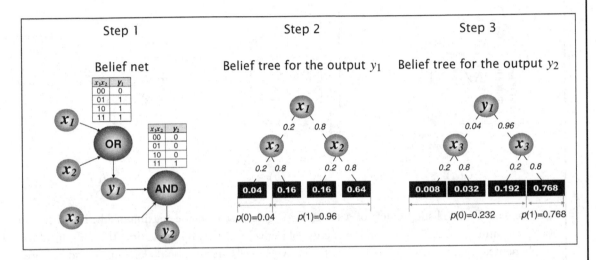

Figure 1.6: A belief tree for the output y_2 for the three-input single-output OR-AND network given in Fig. 1.5.

random variables. In [100], an MRF model of a logic gate is proposed, which uses reinforcement and updating via feedback: the most probably correct states are reinforced, and the error of incorrect (less probable) outputs is reduced. CMOS implementation of a modified MRF model has been reported in [145]. Specifically, reinforcement is understood as a form of positive feedback, which incorporates feedback from the results of an action; that is, the more rewarding the task, the greater the probability that it will be selected again. Note that reinforcement learning is a well known paradigm in artificial intelligence. A particular form of reinforcement learning is based on negative feedback, and is called error-correction learning. In this approach, with every iteration, the amount of error is reduced [40].

1.2.6 NEUROMORPHIC MODEL

In [75, 135, 150], models of logic gates based on Hopfield models with Boltzmann updating rules were considered. Note that the concept of a Boltzmann machine is a probabilistic generalization of

Hopfield models [40]. The Hopfield model consists of a set of computing cells (threshold elements or neurons) and a corresponding set of unit delays, forming a multiple-loop feedback system (Fig. 1.7b). The output of each cell is fed back via a unit delay element to each of the other cells in the model. A logic function is embedded in the Hopfield model using the compatibility truth table (function) whose minterms correspond to the states of the Hopfield model and energy function. Depending on the signal-to-noise ratio, dozens of iterations are required, in order to ensure the stability of these networks with respect to the corresponding ground states and outputs. There are various ways to accelerate computation for specific switching functions.

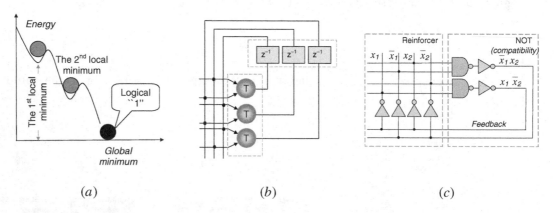

(a) (b) (c)

Figure 1.7: Probabilistic models of a logic NOT gate based on the maximum-likelihood principle: (a) Computing paradigm based on the energy (objective) function minimization, (b) a Hopfield model implemented by a three-loop feedback network, where cells are threshold elements T and z^{-1} are delay elements [75, 135], and (c) An MRF model implemented by a two-loop feedback PLA based on reinforcement criterion [100].

1.3   HARDWARE IMPLEMENTATION

In the presence of noise, the results of computation carry a high risk of distortion. Sub-threshold (lower power) logic devices such as elementary logic NOT, NOR, and NAND gates use margin reduction techniques to cope with uncertainties in the form of noise, caused by thermal effects, electromagnetic coupling, hot-electron effects, and threshold variations. These uncertainties can thereby be suppressed or mitigated.

Design taxonomy.   The common idea behind Boltzmann (Hopfield) machines and MRF models of logic gates is a well-established procedure for statistical estimation based on maximization of the joint probability distribution (or mutual information); that is, the *maximum-likelihood principle* of computation. Although being similar in final values of estimations, the MRF model and Boltzmann machine lead to similar final estimated values, they are different in implementation. In general, the use of the maximum-likelihood principle offers a number of useful properties and capabilities, such as (a) adaptivity (a built-in capability to adapt designed functionality to changes in the surrounding environment), (b) noise tolerance, and (c) fault tolerance [40].

Logic networks with feedback. Systems with feedback offer a consistent way to deal with uncertainties in the form of noise. The basics of formal analysis of logic networks with feedback were developed between 1950-1960; in particular, by D. Huffman, G. Mealy, E. McCluskey, and E. Moore. Their results are introduced in any textbook on logic network design. As early as 1970, Kautz demonstrated the existence of circuits for which he was able to prove that the minimal form must have cycles [61]. In 1971, Huffman [47] discussed feedback in linear threshold networks. Fredkin and Toffoli defined closed and open conservative-logic circuits using a library of reversible gates with loops as a time-discrete dynamic system, and provided their interpretation in terms of sequential networks [29].

Feedback in switching theory is a concept, firmly associated with sequential circuits; that is, circuits with memory. Examples of binary elements and systems with feedbacks include:

(*a*) An inverter (NOT gate) with a loop is the simplest bistable memory element (without control). Its behavioral characteristics can be changed by serial connections or coupling of inverters.

(*b*) A logic gate with a loop is the simplest bistable memory element with control; such examples include latches and flip-flops.

(*c*) A threshold cell with a loop is the simplest bistable memory element with weighted inputs and threshold output. These are used in the design of neuromorphic models such as associative memory and Hopfield and Boltzmann models of logic noise-tolerant gates [44, 135].

(*d*) A multiplexer with a loop is also the simplest bistable memory element with control, enabling design of loopy binary decision trees and diagrams.

(*e*) PLA with a loop is used to design multi-state feedback systems, based on probabilistic models such as the Markov random field model [98, 99, 145].

A formal concept of feedback in a circuit with memory is known as finite state machine.

Noise-immunity of circuits with feedback. The unique feature of probabilistic iterative models is that noise-immunity becomes a natural property of the model. The role of feedback in such models is two-fold: (a) for state memorization, and (b) for noise reduction. It is well understood in signal processing that feedback in a system with noise mitigates the effect of noise generated inside the feedback loop, reduces the sensitivity of a system to variation in the values of parameters, and reduces nonlinear distortion [40]. Probabilistic iterative models are often characterized by the rate of convergence; that is, the number of iterations required for the model, in response to inputs, to converge "close enough" to the stable state given a chosen criterion of stability (optimum).

The effects of feedback can be demonstrated using the simple linear system with noise given in Fig. 1.8. It may (a) mitigate the effect of noise generated inside the feedback loop by a factor

$1 + GH$, and (b) reduce the sensitivity of a system to variation in the values of parameters; e.g., reduce nonlinear distortion by a factor $1 + GH$ (details can be found in [40]).

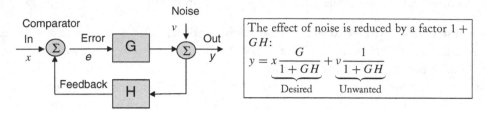

Figure 1.8: A system with feedback and a noise signal incorporated inside the loop.

1.4   CONCLUDING REMARKS

Noise mitigation and noise tolerance are two distinct methods and techniques for handling noise. The noise mitigation approach involves developing noise-analysis tools for detection and identification of hot spots and methods for their mitigation. This approach is fundamentally inefficient in logic circuit design because of its high hardware complexity and energy consumption. In contrast, circuits and systems based on the noise tolerance computing paradigm are inherently tolerant to noise and errors. This approach is preferred in design tools for advanced technologies of the foreseeable future.

In a general sense, the term "stochastic computing" means that various sources of randomness are exploited in computing. In this chapter, we gave a brief introduction to stochastic computing. Probabilistic models provide a convenient platform for studying the trade-offs between the performance of logic gates and their noise tolerance. Enhancing noise tolerance in a logic gate carries various penalties, in particular, increasing delay characteristics, number of iterations, and power consumption. A designer of noise tolerant devices needs an effective technique for minimizing the cost of these penalties given a required level of noise tolerance. We focus on probabilistic models, which are described by joint probability distributions, and can be viewed as obtaining missing information about some object using knowledge (available data) of other information about that object.

1. The simplest way to deal with randomness is averaging, in order to obtain parameters such as the mean and variance. The Gaines' adder and multiplier are based on an implementation of a mean operator via logic AND processing of random binary streams [34]. Stochastic computing, proposed in 1969 by Gaines, is a well-established noise-tolerant computing paradigm which is still popular for the design of reliable low-precision adders and multiplies [15, 50]. Unlike binary radix arithmetic, "stochastic arithmetic" is robust in the presence of noise and single-bit faults, and its accuracy may be controlled using the dimension of time. The Gaines technique can be viewed as *modeling of noise-tolerant and low-precision arithmetic operations*.

2. Bayesian networks are tools for *probabilistic modeling of objects* in the presence of noise, such as large logic circuits without loops. These techniques are based on an efficient local message

passing protocol using factorization of global joint probability distribution functions [52, 104]. An important concept, used in this chapter, is known as an equivalence of the approximate inference in graphs with cycles (loopy belief propagation). The other relevant concept is a type of iterative processing known as turbo decoding [81].

3. Not all probabilistic models are acceptable for hardware implementation of noise-tolerant logic gates. One such model is the Hopfield model. In 1982, Hopfield showed that a network of symmetrically coupled binary threshold cells has a simple quadratic energy function that governs its dynamic behavior [44]. When the cells are deterministically updated one at time, the network settles to an energy minimum. It was suggested in [18] to use this property for modeling logic gates and simple logic circuits. Hinton and Sejnowski found that (a) the energy function can be viewed as an indirect way of defining a probability distribution over all the binary states of the cells, and (b) if proper stochastic updating is used, the Hopfield network produces samples from Gibbs or Boltzmann joint probability distributions [42]; such a Hopfield model is called a Boltzmann machine.

4. Another model that satisfies the requirements of low-cost hardware implementation of noise-tolerant logic gates, is the MRF model. A useful feature of this model is its ability to represent, in a unified manner, different types of uncertainty using a multidimensional Gibbs distribution.

5. The Hopfield model and Boltzmann machine (also known as neuromorphic model) and the MRF model are characterized by (a) the same mechanism of embedding logic functions in the model, and (b) the maximum-likelihood computing strategy, or Gibbs sampler. Their combination constitutes an efficient hardware implementation that uses an iterative adaptive scheme via a multiple feedback loop network.

6. A highly efficient hardware design taxonomy for iterative error correction codes [11, 31] has been utilized in practice for a long time. This technique can be adopted to the design of error-tolerant computing systems [121, 140]. Also, attempts have been made to systematize and extend the known probabilistic techniques for noise- and error-immune designs. In particular, a technique called *computation as estimation* was introduced in [97]; the idea was to treat hardware errors as computational noise.

CHAPTER 2

Nanoscale circuits and fluctuation problems

Murphy's Law

Good judgment comes from bad experience. Experience comes from bad judgment.

In this chapter, basic technologies for fabrication of nanoscale logic circuits and fluctuation problems in these circuits are described.

2.1 NANOSTRUCTURES FOR LOGIC CIRCUITS

In this section, nanostructure and related technologies for the physical implementation of logic devices and circuits are described. The main examples discussed are of III-V compound semiconductor-based nanotechnology. So far, Si-based technology has tended to be used for practical CMOS logic LSICs. Meanwhile, many quantum and single-electron devices have been demonstrated using III-V semiconductors. They have superior electron transport characteristics. The small effective mass of the electron results in large energy quantization, while the long mean free path provides for ballistic transport in the system, in which electrons keep their wave nature.

2.1.1 WHY NANOSTRUCTURES?

The scaling of microelectronics down to nanoelectronics is the inevitable result of technological evolution (Fig. 2.1). The most general classification of the trends in technology is based on grouping computers into *generations*. Using this criterion, five generations of computers are distinguished; each computer generation is 8 to 10 years in length. The scaling of microelectronics down to nanoelectronics is follows: The size of an atom is approximately 10^{-10} m; atoms are composed of subatomic particles, e.g., protons, neutrons, and electrons; protons and neutrons form the nucleus, with a diameter of approximately 10^{-15} m. Note that 2D molecular assembly (1 nm), 3D functional nanoICs topology with doped carbon molecules ($2 \times 2 \times 2$ nm), 3D nanobioICs (10 nm), and *E.coli* bacteria (2 mm) and ants (5 mm) have complex and high-performance integrated nanobiocircuitry [74].

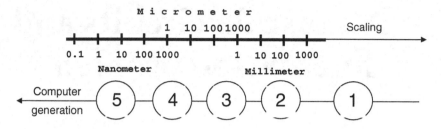

Figure 2.1: Progress from micro- to nanosize in computing devices.

Four attractive features.   From the viewpoint of logic circuits, nanostructures have four main attractive features: (a) High-density integration, (b) High-speed operation, (c) Low-power consumption, and (d) New functionality. Size is directly related to varying performances among electron devices. In the case of conventional Metal-oxide-semiconductor field-effect transistor (MOSFET), it is known as the Dennard scaling principle or scaling law [22]. The law is an important example of Moore's law, empirically indicating exponential growth in the number of devices contained on an LSIC as a function of time [87]. The basic idea behind this scaling is simple. Consider the capacitance shown in Fig. 2.2, representing the gate portion of a MOSFET.

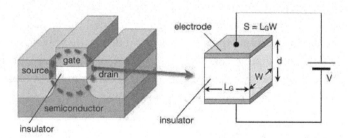

Figure 2.2: Conventional MOSFET structure and model capacitor of gate portion.

Capacitance, switching energy, and energy dissipation.   The capacitance, with insulator thickness, d, electrode area, S, and dielectric constant, ε, is given by

$$C = \varepsilon \frac{S}{d}.$$

(2.1)

The switching energy of the MOSFET is the energy required to charge and discharge the capacitance under the applied voltage, V, and is thus given by

$$E = CV^2. \tag{2.2}$$

Assuming a typical system dimension, L, the dimensions of S and d are represented by L^2 and L, respectively. Then the capacitance reduces to $C = \varepsilon L$ and the energy is found to be proportional to the size,

$$E = \varepsilon L V^2 \propto L. \tag{2.3}$$

Thus, the energy dissipation of the device decreases as its size decreases, resulting in decreasing its power consumption. The ultra-low power consumption of the single-electron devices should be understood in this context. Using the relation $Q = CV$, equation 2.2 can be expressed as $E = QV$. When electrons are the charge for switching, Q is equal to eN, where e is the elementary charge and N is the number of electrons contributing to the switching, then, the switching energy is given by

$$E = NeV. \tag{2.4}$$

This energy is proportional to the number of electrons. In the single electron device, $N = 1$ and the minimum switching energy is obtained. Equation 2.2 also suggests that the potential impact due to voltage reduction is larger than due to capacitance decrease through size reduction.

Switching speed, charging, and traveling time. The switching speed depends on the size, and is characterized by charging time and traveling time. The longer one dominates the speed. The charging time is given by CV/I, where I is current. It becomes small as the size decreases, since the capacitance decreases. The traveling time is evaluated by L/v, where v is velocity of the carrier. Thus, the small size results in the short traveling time.

New functionality. New functions have been expected from quantum mechanical effects and the single-electron effect over the last decade, but most of them work only at low temperatures, because thermal fluctuation smears out their delicate behaviors. However, other new and useful functionalities in nanospace have been discovered recently, as nanotechnology has developed [93, 94, 101, 149].

2.1.2 NANOSTRUCTURE FORMATION

There are two nanostructure formation techniques: the top-down process and the bottom-up process. Fig. 2.3 summarize the techniques.

Top-down formation technique. The top-down process is the way in which a structure is formed using lithography (patterning), thin film deposition, and etching (Fig. 2.4). It has been used for the industrial production of many semiconductor devices and circuits including LSICs. The advanced

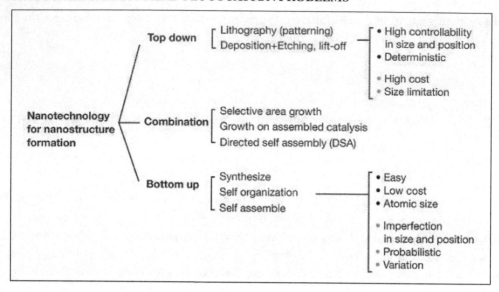

Figure 2.3: Nanotechnologies for nanostructure formation.

fabrication processes of 2012 use 22 nm technology. There are currently several lithography techniques, including optical (photo) lithograph, electron beam (EB) lithography, nano-imprints, among others. Optical lithography has been the mainstay of industrial mass production. The resolution of optical lithography is almost limited by the wavelength of the optical source used. Currently, an ArF laser, generating 193 nm-wavelength light, is used. There are several techniques for making fine patterns with dimensions less than the wavelength, such as liquid immersion [131] and double patterning [25]. In order to follow the outline of the international technology roadmap for semiconductors (ITRS) [46], further smaller patterning using extremely ultra-violet light (13.5 nm) and X-rays is under development. However, the costs of the lithography machines and masks are now extremely high, creating quite a serious problem for this technology. EB lithography can realize patterning as fine as 10 nm. However, the pattern is generated with the direct writing of the fine beam, and the throughput is very low. Although this technique can be used for the special purpose such as researches, it is difficult to apply it to the mass production at present.

Bottom-up formation technique. The bottom-up process produces the various nanostructures by taking advantage of the self-organized nature of the material. Examples of this approach include the fabrication of InAs dots by Stranski-Krastanov (SK) growth [56], nanowires by VLS (vapor-liquid-solid) growth [56, 57], and carbon nanotube [48] among others. This technique has been developed considerably in the last decade, and plays an important role in nanotechnology. It can produce nanostructures en masse in a simple way and at relatively low cost. However, the precise control of position and size is difficult and this technique does not make a good fit to the conven-

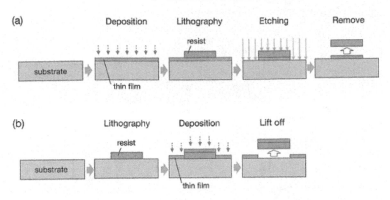

Figure 2.4: Top-down process: (a) etching, and (b) lift-off processes.

tional semiconductor LSI technology. On the other hand, nanocomposite and bundled structures frequently possess unique electronic and/or optical properties that are not seen in bulk materials. CNT sheets [148] and nanomesh metal sheets [96] can be used as transparent conductive sheets, for example.

Mixed top-down and bottom-up processes.   Recently, technology combining the top-down and bottom-up processes has been developed. Basic position and size is defined in the top-down fashion, while fine structures are formed in the self-organized fashion (Fig. 2.5); for example, semiconductor selective area crystal growth through a patterned mask on the substrate [32], CNT growth on size-controlled nanoparticles as catalysis [117], and so on. Direct self assembly (DSA) is an emerging technology for nanoscale patterning. Basic resist patterns are formed by the conventional technique, and further fine patterns are generated by the self-organizing behavior of block copolymers [129].

2.1.3   NANOSTRUCTURE NETWORK AND SWITCHING FUNCTION FOR CIRCUITRY

The nanostructure networks are very attractive for hardware implementation of logic functions. They are also strongly related to the representation of logic functions. Using Shannon expansion, any logic function can be mapped onto a binary tree, in which each node has one input branch (edge) and two output branches. The goal is to topologically map the graph on a nanowire network. Recent developments in nanotechnology give us the opportunity to create a variety of network structures by controlling materials, formation techniques, and formation conditions. Examples of possible logical structures and corresponding nanowire networks structures are summarized in Fig. 2.6. Details of embedding computational properties into various topological structures are given in [125, 154].

The III-V compound semiconductor networks.   Recently, III-V compound semiconductor networks utilizing the top-down process have been developed for demonstration of the nanostructure-

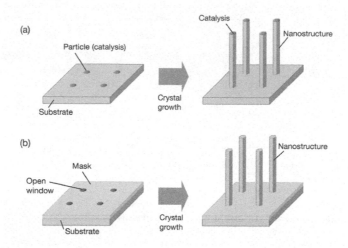

Figure 2.5: Combination of top-down and bottom-up processes: (a) VLS with catalysis particles and (b) selective area growth on patterned mask.

based logic circuits [58, 60, 161]. Here, this technology is explained as an example. As a host material, an AlGaAs/GaAs modulation doped structure grown on a semi-insulating GaAs substrate is used. The layer structure and energy band diagram is shown in Fig. 2.7. The structure is a single crystal grown by molecular beam epitaxy (MBE) or metal-organic vaper phase epitaxy (MOVPE). Only the top AlGaAs layer is doped with n-type donors. It works as either a barrier layer or a carrier supply layer. Electrons moves from the AlGaAs to the GaAs channel layer, since the energy of electrons in AlGaAs is higher than in GaAs. Thus, electrons can be transported in the GaAs channel without impurities, and this system shows superb electron-transport properties. A spacer layer is usually inserted between the top barrier and the GaAs channel in order to prevent unintentional diffusion of dopant atoms from the n-AlGaAs to the channel. The electrons form dipoles with the ionized donors in the AlGaAs and accumulated in the AlGaAs/GaAs interface. Then the energy band bends, as shown in Fig. 2.7(b), and the motion of electrons is limited in the interface plane. These electrons are usually called as two-dimensional electron gas (2DEG). The nanowire network structure is formed on this material using the top-down process. A resist is coated on the material, and network patterns are written by electron-beam lithography and are developed. The network structure is defined by the wet chemical etching, using the patterned resist as a mask. The direction of nanowires must be carefully adjusted along the low index plane of the substrates, such as [-110] and [110] on a (001) substrate, for example. Then, the sidewalls of the nanowire are formed with stable crystal faces with an atomically smooth surface, which is very important for forming good quality gates for field effects on the nanowire. Such sidewalls can be obtained despite the line edge

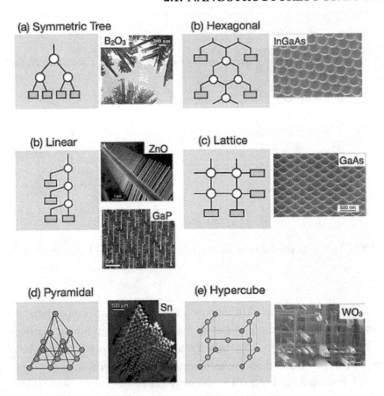

Figure 2.6: Logical and physical network structures for implementation of BDD (reprinted from (a) [73] with permission from American Chemical Society, (b) [24] with permission from Nature Publishing Group and [151] with permission from American Chemical Society, (c) [69] with permission from Elsevier, (d) [95] with permission from American Chemical Society, and (e) [162] with permission from John Wiley and Sons).

roughness of the resist pattern, because they tend to be formed with crystallographically stable planes. Note that the details of the side facet depend on the nanowire material, direction, and etchant [1].

Logic operations.   The logic operations in these lectures are mainly based on the binary switches. A *switch* is any device by means of which two or more conductors (electrical, electrochemical) can be conveniently connected or disconnected. The status of the contact, which can be opened or closed, can be represented by a variable x_i. A switch is the simplest computing device that operates on a single bit. Switches are the basic elements for construction of logic gates. For example, a NOT gate can be implemented as a single-bit switch. Decision trees and diagrams are the graphical data structures, which correspond to the networks of switches. Therefore, they are called *switch-based computing structures*.

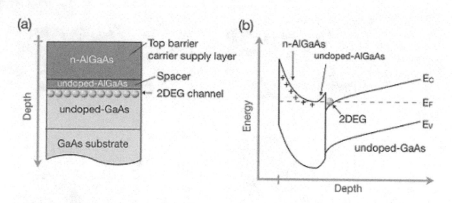

Figure 2.7: AlGaAs/GaAs modulation doped heterostructure: (a) Layer structure and (b) band diagram.

To operate the network as switch arrays, it is necessary to implement a path switch function on each network node. An example of the decision diagram circuit on nanowire networks is shown in Fig. 2.8 [159, 161]. In case of GaAs-based materials, Schottky gates can be obtained easily, where a metal electrode directly formed on the semiconductor nanowire works as a gate [59]. This structure operates as a conventional field effect transistor (FET).

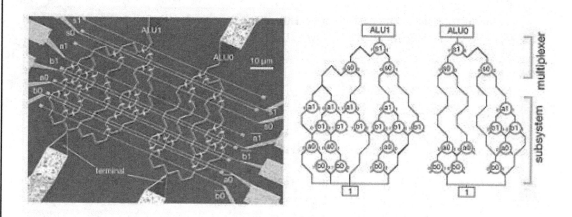

Figure 2.8: (a) Secondary electron microscope (SEM) image, and (b) circuit diagram of a two-bit ALU on GaAs-based nanowire network.

Embedding decision diagram into lattice structure. An embedding of a *guest* graph G into a *host* graph H is a one-to-one mapping $\varphi: V(G) \to V(H)$, along with a mapping α that maps an edge $(u, v) \in E(G)$ to a path between $\varphi(u)$ and $\varphi(v)$ in H.

A switching function f is *symmetric* in variables x_i and x_j if $f(x_i = 1, x_j = 0) = f(x_i = 0, x_j = 1)$, that is, a totally symmetric switching function is unchanged by any permutation of its variables. For a totally symmetric function of n variables, the reduced truth vector has $n + 1$ elements. For example, $f = x_1 \oplus x_2$ and $x_1 x_2 \vee x_2 x_3 \vee x_1 x_3$ are totally symmetric functions. Function $f = x_1 x_2 \vee x_3$ is not a totally symmetric function but it is symmetric with respect to x_1 and x_2. In Fig. 2.9, the symmetry in variables x_i and x_j is utilized in the decision tree and diagram construction, where the node S implements Shannon expansion.

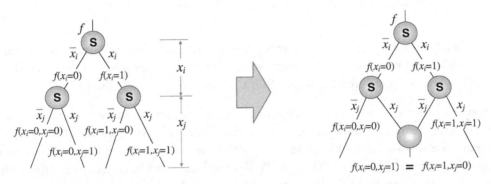

Figure 2.9: Fragment of decision tree for a totally symmetric switching function with respect to x_i and x_j variables (left) and reduced decision tree (right).

Formally, the Shannon expansion of reduced decision tree results in new node:

$$f = \overline{x}_i \overline{x}_j f_{00} \vee \overbrace{\overline{x}_i x_j f_{01} \vee x_i \overline{x}_j f_{10}}^{Symmetry} \vee x_i x_j f_{11} = \overline{x}_i \overline{x}_j f_{00} \vee (\overline{x}_i x_j \vee x_i \overline{x}_j) f^* \vee x_i x_j f_{11}$$

where $f_{01} = f_{10} = f^*$. While decision diagrams often provide a compact representation of switching functions, their layout is not much simpler than that of "traditionally" designed logic networks, making placement and routing a difficult task. As an alternative, *lattice* Binary Decision Diagram (BDD) (L-BDD) have been developed by Perkowski et al. [106]. Earlier, Sasao and Butler [115] developed a method for embedding decision diagrams into an FPGA topology. The L-BDD is based on the *Akers array* [4], which is defined as a rectangular array of identical cells, each of them being a multiplexer, where every cell obtains signals from two neighbor inputs and gives them to two neighbor outputs. All cells on a diagonal are connected to the same (control) variable. Variables have to be repeated to ensure realizability of a given or an arbitrary single-output completely specified switching function. The embedding problem is formulated as embedding a given BDD (the guest structure) into an appropriate regular lattice structure (the host structure). An example of embedding a decision tree into a lattice is given in Fig. 2.10.

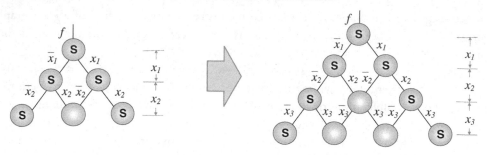

Figure 2.10: Embedding a decision tree of totally symmetric switching function of two variables (left) into a lattice structure as the BDD with extension to three variables (right).

An arbitrary switching function represented by a decision tree can be embedded into a lattice structure using the technique known as *symmetrization* based on *pseudo-symmetry* [14]. This implies that a BDD can be transformed into a L-BDD. Details can be founded in [125].

Path switching.   Path switching is carried out by giving complementary input to the gate formed on each output branch as shown in Fig. 2.11(a). Path switches utilizing quantum and single electron transports are also available. When the gate length is much shorter than the mean free path of the electron ($> 1 \mu$m at 77 K and < 100 nm at 300 K for the AlGaAs/GaAs system) and the confinement energy is much larger than kT (k is Boltzmann constant, T is temperature) in the narrow channel, conductance quantization appears [159]. The transfer curve shows abrupt steps whose height is defined by $2e^2/h$. Then, switching can be accomplished with ultra small voltage swing, resulting in low-power consumption as seen in equation (2.2). Forming two very narrow gates less than 100 nm on each nanowire branch as shown in Fig. 2.11(b), the branch works as a single-electron switch.

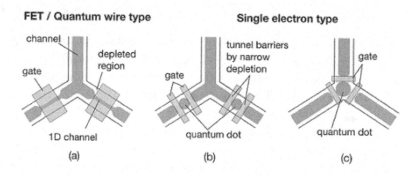

Figure 2.11: Path switching devices on network nodes: (a) FET and quantum wire type, (b) single electron type controlling exit branches, and (c) single electron type with a quantum dot in the node.

Coulomb blockade.   Thin tunnel barriers are formed underneath the gates and a quantum dot is formed in between them. Once an electron is trapped in the dot, it prevents the next electron from entering by Coulomb repulsion and it turns off the current in the branch, in a process called a Coulomb blockade. By attaching another gate just on the dot, the threshold voltage of the switching can be adjusted via the applied voltage. The operation temperature may be roughly estimated using the charging energy, $kT < E_C = e^2/2C_{dot}$, where C_{dot} is the dot capacitance. Usually, E_C is much smaller than the thermal energy at room temperature; thus, the single electron behaviors are easily smeared out. However, E_C increases by decreasing the dot size. Several demonstrations of room temperature operation have been reported [113]. Another type of a single-electron path switch device has also been developed, in which a narrow gate is formed on each branch and a quantum dot is formed at the node (Fig. 2.11(c)). This device also exhibits abrupt path switching, even after supplying the single input to only one output branch gate without the complementary signal [93].

2.2   FLUCTUATION IN NANODEVICES AND THEIR INTEGRATED CIRCUITS

Fluctuation becomes noticeable as the size of the device decreases, and represents a very severe issue. Fluctuation must be eliminated in the deterministic logic circuits, because it causes error. In this section, fluctuations in the electron devices and their integrated circuits are reviewed.

Noise and variation.   In semiconductor electronics, fluctuation is generally classified to include noise and variation. Generally, noise refers to fluctuations in a signal (time domain), and variation refers to time-independent difference in samples (space domain).[1] This classification may depend on the time scale of the phenomena, and on the operation speed that we are concerned with. From the viewpoint of signal engineering, noise simply means any signals, except the one we are interested in. Usually, fluctuation is characterized by a spectrum and a probability distribution. These depend on the physical process and/or rule behind the phenomenon. Recently, time-domain behavior of noise becomes important, as well as that in the frequency domain, since the handled signal is single shot, rather than periodic.

Classification.   There are many kinds of fluctuation in electron devices and circuits (Table 2.1). Usually noise in FETs is composed of $1/f$ noise (Flicker noise) of low frequency, and thermal noise (Johnson-Nyquist noise) of high frequency (Fig. 2.12).

Thermal noise.   Thermal noise is caused by the thermal fluctuation of carriers in a resistor, expressed by $V_n^2 = 4kTR$, where V_n is noise voltage and R is resistance. It shows a flat spectrum in the frequency domain. Thermal noise on capacitance C is referred to as KTC noise, given by $V_n = \sqrt{kT/C}$. $1/f$ noise is ubiquitously seen not only in electron devices, but also in nature. Ex-

---

[1]Generally, the variability as statistical factor is defined as the degree to which values of a variable are dispersed. That is, variability, assesses the degree of dispersion around a measure of central tendency. If most values are tightly clustered, variability is low. If values are widely dispersed, variability is high.

Table 2.1: Fluctuation in electron devices and circuits.

	Time dependence	Phenomena	Origin
Variation	Independent	V_{th} variation I_{DS} saturation current variation; switching speed variation	Dopant position/number variation; Gate metal grain; Poly Si crystalline randomness; Atomic-level size variation
Instability, Fluctuation, Degradation	Dependent	Drain current fluctuation V_{th} (offset) fluctuation; gate leakage current	Carrier trap charging/discharging; material physical damage; chemical reaction; static electric shock
Noise	Dependent	$1/f$ noise; thermal noise; shot noise; clock-harmonic noise; switching noise; ringing; side gating; cross talk; external EM noise	Carrier thermal fluctuation; carrier trap charging/discharging; statistical carrier number fluctuation; cosmic lay; radio active lay; displacement current; impedance mismatch; unintentional carrier injection; EM coupling

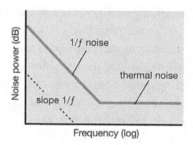

Figure 2.12: Schematic illustration of noise spectrum in a FET.

amples of measured noise waveforms and spectra in GaAs-based nanowire FETs are shown in Fig. 2.13.

Bit error rate. Noise causes an error in logic operation, characterized by bit error rate (BER). BER in this case is evaluated via voltages as follows:

$$\frac{1}{2} \ erfc \ \frac{V_{in}}{\sqrt{2}V_n}, \tag{2.5}$$

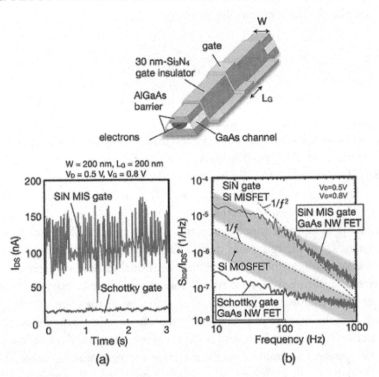

Figure 2.13: (a) Noise waveforms and (b) noise spectra of GaAs-based nanowire FETs.

where *erfc* is a complementary error function. BER (2.5) cannot be 0, since thermal noise is unavoidable. For deterministic operation in a digital logic circuit, the system needs the BER, which is less than 10^{-28}. To achieve this value, a sufficiently large signal input, $V_{in} > 40kT$ (1̃ eV at room temperature), is required. From a physics viewpoint, if N electrons are used to represent a bit signal, thermal fluctuation of each electron is averaged to zero. Therefore, V_n is effectively reduced to $V_n/\sqrt{N}$, causing a decreased BER. Currently, a bit signal is represented by thousands of electrons, resulting in a small enough BER. However, if N will be reduced, in future, to decrease power consumption, BER will increase, as follows from equation 2.4, and the operation will become stochastic.

$1/f$ **noise.** In the case of FET, the most plausible model of $1/f$ noise is the charging and discharging of the traps in the gate region. It is known that single trap charging/discharging events show random telegraph signal (RTS) noise [21, 63]. Its spectrum is Lorenz type and its slope is $1/f^2$. $1/f$ spectrum is attributed by the combination of $1/f^2$ spectra having various time constants [63]. Usually, the following formula is used for expressing low frequency noise, in source-drain current

I_{DS}, represented by the drain current noise power,

$$S_{I_{DS}} = \frac{e^2 N_t g_m^2}{C_{G0}^2 W L_G f^{A_f}} \tag{2.6}$$

where N_t is the trap density, g_m is the transconductance, C_{G0} is the gate capacitance per unit area, W is the channel width, L_G is the gate length, and A_F is the flicker noise exponent. The current noise is converted to voltage noise through the load. Experimentally, A_F takes values between 1 and 2. An intermediate value may be caused by a combination of traps having appropriate time constants [89]. The charging and discharging of the traps causes fluctuations in carrier density or mobility [43], resulting in noise. From equation (2.6), it is understood that the noise increases as the size, W and L_G, decreases. When the device size becomes comparable to the inverse of the trap density, it contains a countable number of traps. At this point, RTS noise dominates, and the $1/f^2$ spectrum appears. If the charging and discharging time constant is longer than the signal period, it is regarded as variation, namely offset voltage shift or threshold voltage shift. It is predicted that the nanodevice is suffering from shot noise, since the number of carriers is decreased. This noise also originates from the discrete nature of electric charge in the channel, which is proportional to $1/\sqrt{N}$ (where N is the number of electrons). In the case of a quantum device, thermal fluctuation is a severe problem. This restricts its operation at room temperature. Even at low temperature, various forms of noise appear. For example, each scattering event appears as noise. These do not appear in macroscopic devices, since each event is averaged out due to the presence of many carriers. In a quantum device, however, current noise pattern sometimes reproducibly appears, and its amplitude is related to the value of quantized conductance. This is called universal conductance fluctuation (UCF) [70]. Reflection and interference of electron wave functions also appears as variation in the output signal. In the case of degenerated states close to each other in space, the electron oscillates, hopping back and forth between the states, resulting in noise. These are sensitive phenomena and are easily smeared out if temperature increases [138].

Power supply noise. Logic circuits also suffered from various forms of noise. Power supply noise and common noise are ubiquitous and severe ones. An example is shown in Fig. 2.14.

These affect all the devices in the circuit. When the FET switches on, current flows from the voltage source to the ground. This event takes place at certain clock intervals in the circuit. Inductance in wires and devices generates the harmonics of the clock signal. Refections of this signal due to the impedance-mismatched devices causes interference. It may be very noticeable in nanodevice integrated circuits, because the handling current is small, and the impedance of the device is very high. The spectrum of this noise is composed of the harmonics of the clock signal. Prima facie, such noise might be thought as deterministic and predictable. However, the large number of integrated devices makes it almost stochastic. Intensive study has been made on noise in the logic circuits using a special technique recently developed [116, 118]. Crosstalk and side-gating phenomena are also a source of unwanted signals and errors.

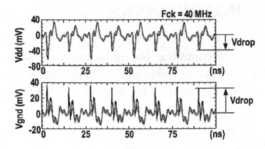

Figure 2.14: Example of on-chip noises in 32-bit linear feedback shift register cells (Reprinted from [92] with permission from IEEE).

Crosstalk and side-gating noise.   Crosstalk takes place between the wires due to electromagnetic coupling. The side-gating in the FET integrated circuit is an unintentional change in drain characteristics, affected by neighboring devices. It is caused by unintentional carrier injection from the neighboring devices in cases of insufficient device isolation. Latch-up in bipolar transistor ICs is similar to it. Side gating is pronounced in III-V semiconductor ICs. Unstable surfaces may induce the phenomenon [53].

Threshold voltage variation.   Variation in gate threshold voltage, V_{th}, is one of the most serious issues in advanced LSICs [136], as it increases as the design dimensions decrease. Currently, variation of a few hundred mV has been observed, which is mainly due to the atomic-level fluctuations in the fabrication process. The degree of variation depends on the size according to $1/L$. Fluctuations in the number of dopants and the individual dopant positions shifts V_{th}. Atomic level differences in the gate insulators of MOSFETs also cause the V_{th} variation and cutoff frequency. At present, such variation is mitigated by large noise margins in the circuit design. However, these large margins require a large logic swing and high supply voltage, which limits the voltage scaling and is an obstacle for reduction of the power consumption.

External noise.   External noise is also a difficult issue for nanoelectronics. The recent increase in wireless communications has filled the space with electric-magnetic waves—all of them potential noise sources. Currently, power electronics has become important for high-efficiency and precise power control in applications such as hybrid cars, electric vehicles, and so on. There are many electric power inverters which change AC to DC and vice versa, and converters which change the voltage of an electrical power source. They perform kHz $\sim$ MHz switching of high voltage and current using power FETs. These emit strong switching noise. Although power electronics is controlled precisely by lower power nanoelectronics, the operation of the latter suffers badly from noise generated by the former. Therefore, noise tolerance becomes an important issue.

2.3   CONCLUDING REMARKS

1. Various nanostructures are formed by top-down and bottom-up nanotechnologies. The top-down techniques are well-controlled, but the process and equipment are extremely complicated and expensive. The bottom-up techniques provide small and high-density nanostructures in a much simpler fashion and at low cost, although the perfect control of size and position of their components are difficult to achieve.

2. An example of realization of nanoscale logic circuits is the III-V compound semiconductor nanowire network and the gate control of the electron transport, presented in this chapter.

3. Nanoscale electronic devices and integrated circuits are characterized by fluctuations, including variation, instability, and noise.

4. Fluctuation is a severe problem in advanced nanoelectronics. Therefore, novel techniques to combat the above problem are necessary. Some of these techniques are described in this book.

CHAPTER 3

Estimators and Metrics

Murphy's proof by contradiction

```
Step 1: State your theorem,
Step 2: Wait for someone to disagree, and
Step 3: Contradict them.
```

In this chapter, we introduce metrics as a system of related measures that facilitates for the quantification of uncertainty and some of the various characteristics of noisy devices. These metrics include probability, joint and marginal distributions of probabilities, information (entropy) measures based on the probabilities of events that convey information, and particular measures traditionally used in signal processing, such as the signal-to-noise ratio, bit-error rate, and Kullback-Leibler divergence.

3.1 WHY DO WE NEED NEW METRICS?

Computing under uncertainty could be called the risky computing, in the sense that the risk of obtaining an incorrect result is higher than it is in the context of conventional computing. Various techniques and models have been developed for the partial minimization of these risks, such as belief networks [52, 105], fuzzy models [111, 130], Dempster-Shafer models [124], decision diagram techniques [130, 154] (in which uncertainty is propagated in the form of "don't care" values), and information-theoretical models [2, 19, 79, 125]. These approaches require special techniques for measuring their efficiency; in particular, error risk assessment, risk constraints, and others characteristics related to risk. Note that there is a difference between the term "risk" and "uncertainty": uncertainty is analyzed for the purpose of measuring risk, which is defined as the probability that an unfavorable event occurs.

3.1.1 OBJECTIVE AND SUBJECTIVE MEASURES OF BELIEF

Uncertainty is understood in the sense that the available information allows for several possible interpretations, and it is not entirely certain which is the correct one. For example, a statement that "the temperature is about seventy degrees" induces a probabilistic uncertainty with respect to the value of the temperature. Probability is a measurement of the degree of uncertainty; reflects a measure of one's belief in the occurrence of an event. Probability is derived from various sources, such as (a) long-run frequency called *objective probability*, or (b) an individual's personal feeling (as expressed in their betting odds) called *subjective probability*.

In the objective concept, probability is the relative frequency in the repeated process; this concept can be applicable only to an event repeated over and over under the same conditions. In the subjective concept, probability is interpreted as a measurement of personal belief in a particular event; it presumes that surrounding conditions or environmental changes in the real world make it practically impossible to maintain the assumption that all trial conditions are identical. Subjective probabilities are used in engineering practice for non-repeatable events and they are typically assigned by expert judgment based on available evidence and previous experience.

3.1.2   LOGIC OPERATIONS AND DATA STRUCTURES

A logic function is specified by its truth table, and can be described in various forms, such as its sum-of-products and polynomial representations, using logical or arithmetic operations [153, 154]. The *arithmetic expression* or *integer-valued representations* of a Boolean function is the bridge between its logic representation (using logic operations) and probabilistic description (using arithmetic operations). For example, given two binary variables x_1 and x_2, the following is true: $\bar{x} = 1 - x$, $x_1 \vee x_2 = x_1 + x_2 - x_1x_2$, $x_1 \wedge x_2 = x_1x_2$, and $x_1 \oplus x_2 = x_1 + x_2 - 2x_1x_2$. In general, a Boolean function of n variables is the mapping $\{0, 1\}^n \rightarrow \{0, 1\}$, while an integer-valued function in arithmetical logic denotes the mapping $\{0, 1\}^n \rightarrow \{0, 1, \ldots, p - 1\}$, where $p > 2$.

These forms of representations are acceptable for small logic functions and are useful for local manipulation. However, in order to handle a large logic network, containing hundreds, thousands or millions of gates, a special graphical data structure, called a *decision diagram*, is used [16, 154]. Various metrics may be used to evaluate the efficiency of this graphical data structure, such as its number of nodes, shortest and longest path, switching activity, delay, and area. However, they only provide a probabilistic interpretation of Shannon expansion (corresponding to each node of the decision diagram) once probabilities are assigned to the paths [154].

3.1.3   OPERATIONS WITH PROBABILITIES AND DATA STRUCTURE

Probabilistic models are designed based on the rules for manipulations of probabilities. A logic function cannot be directly incorporated into a probabilistic model because of their different domains (logical versus arithmetic). Specifically, each logic function must be replaced by its arithmetic analog. Consider the random binary variables x_1 and x_2 with probabilities of being "1" $p(x_1)$ and $p(x_2)$, respectively. Multiplicative rule states that probability of x_1x_2 is $p(x_1x_2) = p(x_1)p(x_2|x_1)$. In particular, the special case where x_1 and x_2 are mutually independent, $p(x_1x_2) = p(x_1)p(x_2)$. Additive rule states that $p(x_1 + x_2) = p(x_1) + p(x_2) - p(x_1)p(x_2)$, and for the special case of mutually exclusive x_1 and x_2, we have $p(x_1 + x_2) = p(x_1) + p(x_2)$. It can be shown that these multiplicative and additive expressions for two random variables (events x_1 and x_2) are the arithmetic forms of the corresponding Boolean functions of the two binary variables x_1 and x_2. Logic function is embedded into the model using appropriate mechanisms, such as its compatibility truth table, arithmetic representation, and an intermediate function called the energy function. These forms of probabilistic description are acceptable for small (local) processing. In probabilistic models, global description is represented by a joint probability distribution function, whose computational complexity is not acceptable for large logic networks (in probabilistic models, the number of logic variables is equal to the number of random variables in the joint probability distribution function). Factorization of the joint probability distribution function, and the technique of employing graphical data structures, both help to overcome the above limitations.

3.1.4   MEASURES OF UNCERTAINTY

The multiplicity of modalities associated with the kinds of information that can appear in computing devices requires various measures for representing uncertainty. Depending on the available information, one may have probability measures, possibility measures, likelihood measures, information-theory measures, fuzzy measures, Demster-Shafer measures, or some combination of the above uncertainty measures [26, 120]. Close connections exist between some of them. Rather than being in competition, these different forms of measure are needed to represent various types of uncertainty, such as randomness, soft errors, and lack of specificity among others. Probabilistic models are designed in terms of beliefs, measured by probabilities and described by probability distribution functions. These models require *statistical performance metrics* such as signal-to-noise ratio (SNR), bit error-rate (BER), and the Kullback-Leibler Distance or Divergence (KLD). Because probability arithmetic operates with real numbers, special techniques are needed to process it at the bit-level.

3.2   UNCERTAINTY REPRESENTATION AND ESTIMATION

Let X be a variable, which attains its value in the space that takes some value within the set $S = \{x_1, x_2, \ldots, x_n\}$. When the exact value of the variable X is unknown, the best we can do is to try to formulate our knowledge about X in a useful mathematical structure. Statistical methods suggest that the exact value of variable X is replaced by our belief about the value of X. The measure of X is interpreted as a measure, associated with our belief that the value of X is contained in the space S; this measure is the *confidence* we have about $X \in S$.

3.2.1   VARIABILITY AND RANDOM VARIABLES

By *variability*, we mean that successive observations of inputs or outputs do not produce exactly the same result. Various factors represent potential sources of variability in the device or system. In models, this variability is referred to as a *random variable*. The term "random" is used to indicate that noise disturbances can change its measured value. A random variable X is used to describe a measurement by the model $X = \mu + \varepsilon$, where μ is a constant and ε is a random disturbance, or noise term. The constant remains the same; say, 0.5 V, but small changes from various sources of variability change the value of ε. The ideal situation occurs when none of these disturbances are present; that is, $\varepsilon = 0$ and $X = \varepsilon = 0.5V$. However, this never happens in practice.

Therefore, a probability is used to quantify the likelihood that a measurement falls within some range of values. The likelihood is quantified by assigning a number from the interval [0, 1]. The probability of a result, which can be interpreted as a *degree of belief* that the result will fall within some specified range. Given two random variables, X and Y, the probability distribution that defines their simultaneous behavior is a *joint* or *multidimensional* probability distribution. The individual probability distribution of a random variable is referred to as its *marginal* probability distribution. For example, if n random variables are joint normal (Gaussian), any subset of them or linear combination of them is also distributed according to a joint normal distribution. For joint normal random variables, independence and uncorrelatedness are equivalent (this is not true in general for non-normal random variables). The fundamental reason for the importance of Gaussian processes for us is that thermal noise in electronic devices can be closely modeled by a Gaussian distribution. Apart from thermal noise, Gaussian processes provide an acceptable approximation for various information sources as well. Note that in probabilistic modeling, the number of logic variables in a logic function that is being modeled, and is equal to the number of random variables in the joint distribution. A *random process* or *stochastic process* is an extension of the concept of a random variable. Information sources can be modeled as random processes. For more details, refer to the textbooks [84].

3.2.2 PARAMETER ESTIMATION

Any statistic used to estimate the value of an unknown parameter θ is called an *estimator* of θ. The observed value of the estimator is called the *estimate*.

Parametric estimation theory is based on the assumption that the parameters of a system are well-defined, but unknown to the observer, who tries to estimate the system and its parameters as well as possible, using the data set. The model is described by a joint probability distribution $f(X|\theta)$, where θ is a parameter. A good estimator is, on average, "close" to the true value θ. Note that for every realization of the data set, a different estimation is produced. Within this framework, the following parameter estimators can be distinguished:

1. *Maximum likelihood* is used when knowledge about the data distribution is available but not about the prior information, $\hat{\theta} = \arg\max_\theta p(x|\theta)$.

2. *Maximum entropy*, used when only the prior information is available,
   $\arg\max - \sum_{i=1}^{n} p(x_i) \log_2 p(x_i)$.

3. *Bayesian*, used when both the prior and likelihood distributions are known. Specifically, it is assumed that (a) initial knowledge about θ is contained in a known prior distribution $p(\theta)$, (b) the distribution $p(x|\theta)$ (likelihood) is known, but the value of the parameter θ is not known exactly, and (c) the rest of knowledge is contained in a set of samples of the random variable.

Usually, these techniques cannot be applied directly because the probability distributions involved cannot be computed in a tractable manner. Various approximate inference techniques are used in iterated models, based on the maximum likelihood and maximum entropy criteria. The focus of our study is Gibbs sampling. The idea is to approximate the true posterior distribution by a simpler distribution. One measure of similarity between the two distributions is the Kullback-Leibler divergence.

3.2.3 PROBABILITY METRICS

Uncertainty (incompleteness, imprecision, vagueness, unreliability) is understood in the sense that the available information allows for several possible interpretations, and it is not entirely certain which is the correct one. Uncertainty is present in the technology, computing process, and measuring process. For example, a statement that the current is 0.5 mA carries with it a *possibilistic* uncertainty with respect to the value of 0.5 mA and some error; for example 0.5 mA ± 0.001. Probability is a measure of uncertainty, and information is associated with probability.

Consider a two-input NAND gate with two mutually independent random binary inputs x_1 and x_2, assuming their probability of being "1" is $p(x_1) = p(x_2) = p$ (Fig. 3.1a). Thus, we may replace the notion of the correct values of x_1 and x_2 by beliefs (measured as probabilities) about the correctness of these values of these variables. We consider these values to be varying in a random manner over the range of the possible voltage levels. The probability of the output being logic "1" is derived from the truth table of the NAND function, and it is described by the conditional probability (Fig. 3.1a): $p(f|x_1, x_2) = (1 - p)p + p(1 - p) + (1 - p) = (1 - p)^2 = 1 - p^2$.

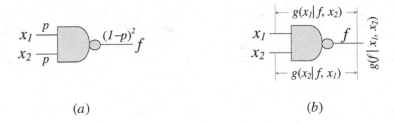

(a)   (b)

Figure 3.1: Probabilistic metric for logic gate NAND with two random inputs x_1 and x_2 (a), and deriving marginal probabilities from joint distribution $g(f, x_1, x_2)$ (b).

Switch models.   Switch models represent various possible physical switch mechanisms, for example, the flow of electrons or other information carriers. A switch is represented by three connection points called terminals to which external signals may be applied or from which internal signals may be drawn. The lines attached to the terminals can represent wires, pipes, or other transmission media appropriate to the particular technology. An input terminal allows a signal to enter the switch and change its state. An output terminal allows signals to leave the switch. The control variable x is applied to an input terminal of the switch.

The effect of a switch on the output is determined by the *state* of the switch. In Fig. 3.2a, the switch has two states: OPEN (OFF) and CLOSED (ON). The former implies that there is no closed path through the switch that connects the upstream and downstream data terminals. In Fig.

3.2b, a switch-based model of a two-input AND gate is shown. Table 3.1 provides a probabilistic description of two-input logic gates based on their switch models.

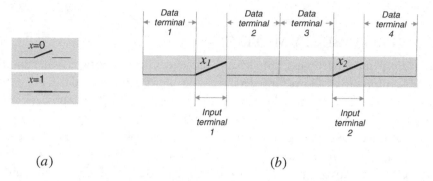

$$(a) \qquad\qquad\qquad\qquad (b)$$

Figure 3.2: Switch-based model of a logic gate: (a) the two states of the switch, OPEN (OFF) and CLOSED(ON), and (b) a model of a two-input AND gate.

Table 3.1: Switch-based models of logic gates.

Function	Logic gate	Truth table	Switch model	Probabilistic model
AND $f = x_1 x_2$	x_1 — x_2 — f	x_1 x_2 f 0 0 0 0 1 0 1 0 0 1 1 1	x_1 x_2	$p(x_1)p(x_2)$
OR $f = x_1 \vee x_2$	x_1 — x_2 — f	x_1 x_2 f 0 0 0 0 1 1 1 0 1 1 1 1	x_1 x_2	$p(x_1) + p(x_2) - p(x_1)p(x_2)$
NAND $f = \overline{x_1 x_2}$	x_1 — x_2 — f	x_1 x_2 f 0 0 1 0 1 1 1 0 1 1 1 0	$\overline{x}_1$ $\overline{x}_2$	$\begin{aligned} & p(\overline{x}_1) + p(\overline{x}_2) - p(\overline{x}_1)p(\overline{x}_2) \\ =\; & 1 - p(x_1)p(x_2) \end{aligned}$
NOR $f = \overline{x_1 \vee x_2}$	x_1 — x_2 — f	x_1 x_2 f 0 0 1 0 1 0 1 0 0 1 1 0	$\overline{x}_1$ $\overline{x}_2$	$\begin{aligned} & p(\overline{x}_1)p(\overline{x}_2) \\ =\; & 1 - p(x_1) - p(x_2) + p(x_1)p(x_2) \end{aligned}$

Probability distribution.   By observing the three random variables involved in a logic gate: inputs x_1 and x_2, and output f (their relationships are being defined by so-called compatibility truth table), their probabilistic behavior is described by conditional distributions $g(x_1|f, x_2)$, $g(x_2|f, x_1)$

and $g(f|x_1, x_2)$, respectively (Fig. 3.1b). In Hopfield and MRF models, the true joint distribution $g(f, x_1, x_2)$ is approximated by a simpler probabilistic distribution called the Gibbs distribution. Given a joint probability distribution $p(x_1, \ldots, x_r)$, the *marginal distribution* is defined as $p(x_1, \ldots, x_s) = \sum_{x_{s+1}, \ldots, x_r} p(x_1, \ldots, x_r)$, $s \leq r$. A marginal distribution can be viewed as a projection of the joint distribution onto a smaller set of variables. That is, global probabilistic characteristics can be computed using local interactions via factorization. In circuits, the effect of noise is modeled by a probability distribution for the nominal voltage or current. A normal distribution is a good approximation for this distribution. However, the propagation of random signals in circuits results in a non-normal distribution.

Factoring joint distributions.   A representation for a high-dimensional (multivariate) joint probability distributions can be factored into a low-dimensional distributions known as a Bayesian network [105], MRF [35], and Boltzmann machine [40]. These models are based on the principle of approximating the true posterior joint distribution by a simpler distribution, such as a Gibbs distribution. One measure of the similarity between two probability distributions is called the Kullback-Leibler divergence [68].

3.2.4   INFORMATION-THEORETIC METRICS

Various models involving probabilistic uncertainty make use of the Shannon entropy as a criteria function for comparing uncertainty-reduction techniques [2, 19, 20, 79]. For example, the principle of maximum entropy suggests that the prior distribution that maximizes entropy shall be selected; this measure is a manifestation of maximum uncertainty.

Communication in computer systems.   Because of data transmission between the gates in a logic network, the techniques of information theory can be used for various design tasks; in particular, in noise-immune and fault-tolerant modeling. A computing system can be considered as a process of communication between computer components. The classical concept of information advocated by Shannon is the basis for this. However, some adjustment must be made to it in order to capture a number of features of the design and processing of a computing system. Shannon information describes the uncertainty eliminated by a measurement. Information, or entropy, measures the average number of binary digits that are needed to encode a given message using a set of symbols, each with a given probability. This information measures the average length of the binary sequences. Note that in physical systems, information, in a certain sense, is a measurable quantity that is independent of the physical medium by which it is conveyed.

Probability distribution, entropy, and information.   Let $p(x_1), p(x_2), \ldots, p(x_n)$ be the probability distribution of a random variable X. The entropy of X is defined as $H(X) = -\sum_{i=1}^{n} p(x_i) \log_2 p(x_i)$, where $-\log_2 p(x_i)$ is the amount of information gained after observing the event $X = x_i$ with probability $p(x_i)$. The entropy $H(X)$ is a measure of the average amount of information conveyed per message. For example, the entropy for the Boolean function f of three variables given by the truth vector $\mathbf{F} = [10111110]^T$, is $H(f) = -0.25 \log 0.25 - 0.75 \log 0.75 = 0.8113$. The entropies of the variables are the same, $H(x_1) = H(x_2) = H(x_3)$, and equal to $-0.5 \log 0.5 - 0.5 \log 0.5 = 1$.

Mutual information. Mutual information is the difference between the entropy and the conditional entropy, $I(f; g) = H(f) - H(f|g)$. This can be seen as a difference between the uncertainty of f and the remaining uncertainty of f after coming to know g. This quantity is the information of f, obtained by knowing g. Mutual information is, thus, a measure of the degree of dependency between f and g. Let $\mathbf{X}$ and $\mathbf{Y}$ be a pair of random vectors. The mutual information, $I(\mathbf{X}; \mathbf{Y})$, can be viewed as a measure of the uncertainty about the output $\mathbf{X}$ of a system that is resolved by observing the system input $\mathbf{Y}$. The maximum mutual information principle is a well-established computational technique which is used in the optimization, in particular, of decision trees and diagrams of logic functions [154]. This principle operates by maximizing the mutual information $I(\mathbf{X}; \mathbf{Y})$ employing a redundancy in the $\mathbf{X}$ compared with $\mathbf{Y}$.

The details of information-theoretical measures are given in [6, 7, 110, 112]. The maximum entropy principle and maximum mutual information principle have been used for computing and optimization of logic functions using decision diagrams. The initial state can be characterized by the entropy $H(f)$ of f, and the final state by the conditional entropy $H(f|x)$. A formal criterion for completing the sub-tree design is $H(f|x) = 0$.

3.3 MEASUREMENT TECHNIQUES

In the experimental study of probabilistic models of logic functions, the following metrics are useful: (a) signal-to-noise ratio (SNR), (b) bit error rate (BER), and (c) Kullback-Leibler divergence (KLD).

3.3.1 KULLBACK-LEIBLER DIVERGENCE

Given a stochastic system with a set of known states, let $p(x)$ and $q(x)$ denote the probabilities that a random variable X is in state x under two different operating conditions.

KLD in terms of probability distributions. The Kullback-Leibler Distance or Divergence (KLD) between the two probability distribution functions $p(x)$ and $q(x)$ is defined as follows [68]:

$$KLD = \sum_{\text{States } x} p(x) \ln \frac{p(x)}{q(x)}, \tag{3.1}$$

where the sum is over all possible states of the random variable X and $q(x)$ plays the role of a reference measure. If the distributions are the same then the KLD is zero; the closer they are, the smaller the value of KLD.

KLD in terms of mutual information. The KLD can be defined in terms of mutual information. Mutual information, $I(\mathbf{X}; \mathbf{Y})$, between $\mathbf{X}$ and $\mathbf{Y}$ is equal to the KLD between the joint probability function $f(\mathbf{X}, \mathbf{Y})$ and the product $f(\mathbf{X}) f(\mathbf{Y})$ of the probability distribution functions $f(\mathbf{X})$ and $f(\mathbf{Y})$ [40].

KLD in experimental study. In our experimental study of probabilistic models, the following KLD equation is used:

$$\text{KLD} = \sum_{\text{States } y} p_y \ln \frac{p_y}{p_Y}. \tag{3.2}$$

where p_y and p_Y are the probability distributions of the noise-free output (ideal discrete signal) and the noisy output (real discrete signal), respectively.

3.3.2 SIGNAL-TO-NOISE RATIO (SNR)

The SNR, measured in decibels (dB), is calculated as

$$\text{SNR} = 10 \log_{10} \frac{\sigma_y^2}{\sigma_e^2} \ (\text{dB})$$

where σ_y^2 and σ_e^2 are the variances of the desired signal y and the noise e, respectively.

3.3.3 BIT ERROR RATE (BER)

The BER is the fraction of information bits in error; it is defined as follows:

$$\text{BER} = \frac{\#\ \text{errors}}{\text{Total}\ \#\ \text{bits}}$$

The number of errors due to signal delay (both rise and fall time) is also considered along with errors due to bit flips while counting the total error in the output.

3.4 MAXIMUM-LIKELIHOOD ESTIMATORS

The method of maximum likelihood may be applied to any estimation problem, if the joint probability distribution of the available set of observed data can be derived. Then the method yields almost all known estimates as special cases. The *maximum-likelihood computing paradigm* is defined as a computational scheme based on maximum-likelihood parameter estimation. This paradigm is widely used in computing devices. For example, in [54], a resilient power management technique was proposed. A taxonomy of noise- and error-tolerant computing system design based on maximum-likelihood techniques is developed in [121].

3.4.1 FORMAL NOTATION

The maximum likelihood technique is considered the best method to obtain a point estimate of a parameter. It was developed in the 1920s by the famous British statistician Sir R. A. Fisher. It evolved from the notion that the most reasonable estimator of a parameter based on sample information is the one that produces the largest probability of obtaining the sample. In maximum likelihood estimation, observations can be viewed as incomplete data (the complete data is understood as observed data and the missing data together).

Observed and hidden parameters.   Given a probabilistic model that depends on observed variables and unobserved latent (hidden) parameters, the maximum likelihood estimate is used to find these hidden parameters. Suppose that for the random variables $X_1, X_2, \ldots, X_n$, their joint distribution is given, except for an unknown parameter θ. For example, the sample could be from a normal distribution having an unknown mean and variance. We are interested in using the observed values to estimate θ. Whereas in probability theory it is customary to suppose that all of the parameters of a distribution are known, the opposite is true in statistics, where a central problem is to use the observed data to make inferences about the unknown parameters. A particular type of estimator is known as the *maximum likelihood* estimator.

Likelihood function.   Let $f(x_1, x_2, \ldots, x_n | \theta)$ be the joint distribution of the random variables $X_1, X_2, \ldots, X_n$. Since $f(x_1, x_2, \ldots, x_n | \theta)$ represents the likelihood that the values $x_1, x_2, \ldots, x_n$ are observed when θ is the true value of the parameter, a reasonable estimate of θ is the value, yielding the largest likelihood of the observed data. In other words, the maximum likelihood estimate $\hat{\theta}$ is defined to be that value of θ, maximizing $f(x_1, x_2, \ldots, x_n | \theta)$, where $x_1, x_2, \ldots, x_n$ are the observed values. The function $f(x_1, x_2, \ldots, x_n | \theta) = L(\theta)$ is referred to as the *likelihood* function of θ. Note that the variable of the likelihood function is θ, not the x_i.

In the discrete case, the quantity $L(\theta)$ is the probability $P(X_1 = x_1, X_2 = x_2, \ldots, X_n = x_n)$, which is the probability of obtaining the sample values $x_1, x_2, \ldots, x_n$ given a particular value of the parameter θ. Therefore, the maximum likelihood estimator of θ is the value of θ that maximizes the probability of occurrence of the sample values. The maximum likelihood can also be used in the cases in which there are several unknown parameters to estimate.

Maximum likelihood estimate.   Formally, the maximum likelihood estimate is $\hat{\theta} = \arg\max_\theta L(\theta)$; that is, the estimate of the unknown parameter vector θ is the value $\hat{\theta}$ that maximizes the likelihood $L(\theta)$. Any algorithm, based on the maximum likelihood principle, iteratively improves an initial estimate $\hat{\theta}^0$ by computing new estimates $\hat{\theta}^0, \hat{\theta}^1, \ldots, \hat{\theta}^m$:

$$\hat{\theta}^{t+1} = \arg\max_\theta L(\theta)$$

where $t = 0, 1, \ldots, m-1$ and $\hat{\theta}^{t+1}$ is the value that maximizes the $L(\theta)$. A solution $\hat{\theta}^{t+1}$ may represent a true *global maximum*, or merely a *local maximum*. For more details, we refer to [27, 84].

3.4.2   THE SIMPLEST ESTIMATOR

The sample mean can be recursively computed by a linear system (Fig. 3.3). After each measurement $x[n]$, we compute an updated estimate of the mean $E[n]$, $E[n] = \frac{1}{n} \sum_{k=0}^{n-1} x[n-k]$. In the maximum likelihood principle, nothing is known *a priori* about the unknown quantity, but there is prior information on the measurement process itself. Thus, the recursively estimates of the mean are up to date in the maximum-likelihood sense, since there is no *a priori* information about $E[n]$.

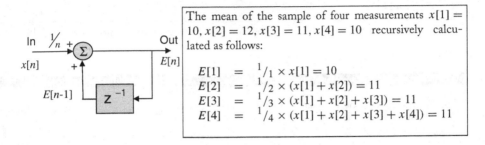

The mean of the sample of four measurements $x[1] = 10, x[2] = 12, x[3] = 11, x[4] = 10$ recursively calculated as follows:

$$E[1] = {}^1/_1 \times x[1] = 10$$
$$E[2] = {}^1/_2 \times (x[1] + x[2]) = 11$$
$$E[3] = {}^1/_3 \times (x[1] + x[2] + x[3]) = 11$$
$$E[4] = {}^1/_4 \times (x[1] + x[2] + x[3] + x[4]) = 11$$

Figure 3.3: The adder with a feedback loop, consisting of a memory unit z^{-1}, which adds the delayed output $E[n-1]$ to the current input $x[n]$ to produce recurrently the mean $E[n]$.

3.4.3   THE BEST HARDWARE ESTIMATOR

An iterative error correcting technique, based on the maximum likelihood decoding principle, is known as *turbo decoding*, and provides error control performance within a few tenths of a dB of the Shannon limit [11]. This bit-level *a posteriori* probability processing exploits the likelihood of specific error magnitudes, in order to correct the output. It has been adopted in various noise- and error-tolerant computing system design taxonomies, in particular, in [97, 121].

Given a received vector $\mathbf{y}$, the maximum likelihood decoder chooses the codeword $\mathbf{c}$, that maximizes the probability $p(\mathbf{y}|\mathbf{c})$. To perform maximum likelihood principle, decoding Hamming distance (for binary signals) or Euclidean distance (for non-binary signals) are used as metrics. The device can produce accurate estimates of $p(c_i|y_i)$, using an iterative scheme with feedback. The maximum number of iterations required by such a decoder varies depending on the level of noise. A decoder is said to have *converged* when further iterations will not change the decoding result. Ideally, decoding is stopped once the decoder has reached its final decision. Various stopping criteria have been proposed, according to which the decoder can recognize when it has run enough iterations. Once decoding has converged, posterior probabilities should agree, and the hypothesis with the highest posterior probability will be favored by the decoder, because this will most likely be the correct one. A useful measure of this fact is the similarity of distributions, such as the KLB.

3.5   SUMMARY AND DISCUSSION

Designers of noise tolerant devices and systems need effective techniques to minimize the cost of designs given a desired level of noise tolerance. This task requires several different metrics. *Estimator* is a general term for various statistical techniques for resulting in computing unknown parameters using the available data. In this book, we focus on a hardware implementation of specific types of estimators. In this chapter, we establish basic metrics for the evaluation of probabilistic computations.

Table 3.2 shows the design differences between the conventional and probabilistic models of a logic circuit.

Table 3.2: Comparison of characteristics for a conventional logic network design and its probabilistic model.

Characteristic	Conventional design	Probabilistic based model
Global description	Boolean function of n binary variables	Joint ($m \ll n$-dimensional) probability distribution
Local description	Boolean function of $m \ll n$ binary variables	Marginal ($m \ll n$-dimensional) distribution
Optimization	Decomposition of Boolean function	Factorization of joint probability distribution
Computation	Logic value propagation	Probability of logic value propagation
Logic circuit design	Design taxonomy	Special mechanism for embedding Boolean functions into the model.

1. The problem of parameter estimation is a classical problem in statistics, and it can be approached in several ways, such as maximum-likelihood estimation and Bayesian estimation. Although the results, obtained using these procedures, are nearly identical, they utilize conceptually different computing paradigms. The references [100, 135, 145, 158] justified the fact that the maximum-likelihood principle is well-suited for hardware implementation of noise-tolerant logic gates and small logic circuits.

2. Risks in computing under uncertainty can be evaluated by various metrics, such as probability, Dempster-Shafer metrics, fuzzy logic metrics, and entropy (information) metrics. In this lecture, we use probability metrics and various related characteristics, such as the probability distribution and the KLB divergence for the purpose of comparison, multidimensional (joint) distribution for the purpose of global description, and marginal probability distributions for the purpose of local processing. In addition, we use the procedures based on averaging parameters, such as the mean, variance, and related SNR metric.

3. Maximum-likelihood methods view parameters as quantities whose values are fixed but unknown. The best estimate of their value is defined to be the one that maximizes the probability of obtaining the samples actually observed. In contrast, Bayesian methods view the parameters

as random variables having some known prior distribution. The observation of sample data converts this to a posterior distribution, thereby revising our opinion about the true values of the parameters.

4. The risks involved in computing under uncertainty can be interpreted under the rubric of *decision-making under uncertainty*, which involves the selection of the best alternative from a set of available alternatives. The worst situation is when there is no information about the underlying mechanism determining the parameters of interest. In this situation, a decision must be made under *ignorance*. In most cases, a device makes a risky decision under uncertainty or ignorance. For example, if a random variable X takes one of two possible states, $X = x_1$ or $X = x_2$, the decision rule is as follows: decide x_1 if $p(x_1) > p(x_2)$; otherwise decide x_2. Hence, instead of making decisions based on the current state of the system, a probabilistic model maintains a set of degrees of belief; that is, a probability distribution over the possible states of the system.

5. Adaptation of the *turbo decode principle*, most efficient maximum-likelihood estimator with respect to information-theoretic criteria, to noise-tolerant logic gate design, is still an open problem.

The above should be considered as an extension of the traditional metrics used in computer device and system design, incorporating the techniques for estimating uncertainty. Noise becomes a factor that can change the design of elementary logic gates and networks. Since there is no way of knowing with certainty the correct logical value ("0" or "1" in binary case) in these gates and networks, we must operate with probabilities of correct values. For example, in computing under uncertainty, the OR operation is $0 \vee 1$ is not equal to 1, but in terms of the probability, $p(0 \vee 1) = p(1)$.

CHAPTER 4

MRF Models of Logic Gates

Murphy's definition of statistical techniques

Statistics is a systematic method of coming to the wrong conclusion with confidence

MRF models of logic gates are useful for probabilistic analysis, as well as acceptable for hardware implementation, due to their superb noise-resistance properties. In order to embed a logic function into the MRF model, the function must be (a) given in the form of its compatibility truth table (instead of a traditional truth table), and (b) represented by an integer-valued expression (instead of a logic one).

4.1 BASIC DEFINITIONS

The fundamental definitions for MRF techniques are based on [23, 31, 35]. We also use the applied results reported in [10, 100, 145, 158]. In terminology, we follow contemporary textbooks on probability and statistic for engineers, such as [84].

Graphical data structures play a crucial role in computational models. They allow for the representation of logic functions and their processing via manipulation of their graphical data structures. For example, in conventional logic design, the particular properties of a logic function are delegated to a directed acyclic graph, called a Decision Tree. Its optimal form is known as a Decision Diagram, which can be directly implemented in hardware. Similar structures are used in probabilistic models of a logic function. In particular, in the MRF approach, the properties of a logic function are embedded into its model using the intermediate (energy) function, compatibility function, and an indirect graph, specified by a set of clique configurations. Processing of large functions (global computation) is accomplished using local computation, which is based on a factorization of a multidimensional probability distribution.

4.1.1 GRAPHICAL DATA STRUCTURE

Let $X = (x_1, x_2, \ldots, x_n)$ be a set of random variables. Each variable, x_i, $i = 1, 2, \ldots n$, takes a finite set of states, for example, two states, $x_i \in \{0, 1\}$, assuming equal probabilities, $p(0) = p(1)$ (binary variables), or three states $x_i \in \{0, 1, 2\}$, assuming $p(0) = p(1) = p(2)$ (ternary variables). Assume that logic "0" and "1" are understood as the random variations of voltage or current around the values corresponding to "0" and "1." For example, both the input and output of a binary NOT gate can be treated as random signals. Consider an undirected graph in which each node corresponds to a random variable x_i. The case of a binary NOT gate is the simplest graphical MRF model with two random variables represented by a two-node graph. The graphs of the MRF for the logic function NOT ($x_2 = \overline{x_1}$) and EXOR function ($x_3 = x_1 \oplus x_2$), along with these functions' compatibility truth tables, are shown in Fig. 4.1, where the valid state combinations correspond to the value "1" of the compatibility function or compatibility vector (**U**), and invalid states are represented by the value "0," otherwise.

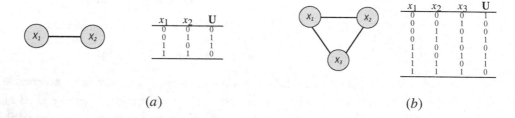

x_1	x_2	**U**
0	0	0
0	1	1
1	0	1
1	1	0

x_1	x_2	x_3	**U**
0	0	0	1
0	0	1	0
0	1	0	0
0	1	1	1
1	0	0	0
1	0	1	1
1	1	0	1
1	1	1	0

(a)   (b)

Figure 4.1: The MRF graphical data structure (network), as undirected graph, and compatibility truth table for NOT gate (a), and two-input EXOR gate (b).

4.1.2   FORMAL NOTION OF THE MRF MODEL

A model is a representation of the device or system in terms of a data structure and/or program. The model can be studied by a process usually referred to as simulation. MRF models of logic gates are acceptable for hardware implementation, and also convenient for the probabilistic analysis of noisy logic gates and networks. The MRF model is also called the *Gibbs sampler*, and is defined as an *iterative adaptive scheme* [40], which generates a MRF with the Gibbs distribution. The Gaussian distribution is a special case of the Gibbs distribution family. Let X be a set of random variables, and G be a *neighborhood*, defined in terms of undirected subgraphs, called a *maximal clique* ("a clique" for short). A clique is a fully connected subgraph. A maximal clique is defined as a clique that cannot be made larger, while still remaining a clique. Examples of the simplest cliques (with two and three nodes) are given in Fig. 4.1. There is a correspondence between a *Gibbs joint probability distribution G*, relative to X, G, and a stochastic process, called the MRF of a random variable X with a neighborhood.

The MRF in log-linear form.   The log-linear form of the MRF is defined as follows:

$$p(X = x) = \frac{1}{Z} \exp \frac{E(x)}{kT} \tag{4.1}$$

$$E(x) = \sum_{c \in C} V_c(x_c) \tag{4.2}$$

where Z is a scaling factor (to normalize the total probability to 1), kT is a control parameter, and $E(x)$ is an *energy* function, which is commonly given by a sum of *clique potentials*, $V_c(x_c)$, over all possible cliques C. The value of $V_c(x_c)$ depends on the local configuration of the clique c.

The MRF in factored form. The factored form of the MRF is defined by the product of all the clique potential functions, divided by a normalizing constant Z,

$$P(X = x) = \frac{1}{Z} \prod_{c \in C} V_c(x_c) \tag{4.3}$$

This form of the MRF (Equation 4.3) is provided by the Hammersley-Clifford theorem (the equivalence between an MRF and a Gibbs distribution) [13, 35]. Using this relationship, the joint probability distribution is factored into terms, each of which depends only on the results of local processing, specified by a set of cliques.

4.1.3 LOGIC FUNCTION EMBEDDING

The embedding of a logic function into an MRF model is accomplished by using

(*a*) Local graphical data structure; that is, neighboring is specified by subgraphs (cliques) with specific properties; and

(*b*) Clique potential functions, $V_c(x_c)$.

The embedding mechanism delegates the logic properties to the model—it "profiles" the model for probabilistic computing of this particular logic function. A logic function must be represented in the form of a *compatibility truth table* (instead of the traditional truth table), or a compatibility function; the latter is defined as any algebraic form (as opposed to logic form) of the compatibility truth table; for example, arithmetic sum-of-product (SOP), arithmetic polynomial, Walsh, or Haar form [125]. In the simplest case of the inverter and various two-input logic gates, the compatibility function represents the function's clique potential. For example, the local properties of the NOT gate are specified by the simplest clique, such as an undirected graph with two nodes, and the energy function is equal to the clique potential: $E(x) = V_c(x_c)$. Any integer-valued form is acceptable for embedding a logic function into this model:

$$E(x) \quad = \quad V_c(x_c) = \underbrace{\overline{x}_1 x_2 + x_1 \overline{x}_2}_{\text{Arithmetic SOP}} = \underbrace{x_1 + x_2 - 2x_1 x_2}_{\text{Arithmetic polynomial}} = \underbrace{{}^1\!/_4 \times [2 - 2(-1)^{x_1 + x_2}]}_{\text{Walsh polynomial}}$$

The above integer-valued forms are equivalent; however, for hardware implementation with feedback (as shown later in this chapter), the arithmetic SOP expression is preferable.

4.1.4 IMPLEMENTATION

MRF models are used for probabilistic analysis of the corresponding hardware implementation.

MRF-based software models. The probabilistic analysis of a logic gate or a small network is provided by computing a joint probability distribution, or marginal distributions (4.1). This computation can be performed directly using the joint distribution. However, a high-dimensional joint distribution can be factored into a low-dimensional distribution (4.3).

MRF-based hardware models. The *arithmetic Sum-of-Products* (SOP) form of a Boolean function is preferable for hardware implementation. A maximum-likelihood computing strategy is implemented by minimizing the energy function. The energy function, which is defined as an arithmetic sum of minterms. This mechanism for Boolean functions is developed in [100]. The idea is that searching for the correct solution (the minimum of the energy function) is implemented by reinforcement, a well known computing paradigm in artificial intelligence and circuits with feedback [40]. For this, a device called a *reinforcer* was proposed in [100], in the form of a logic array with multiple feedbacks.

4.2 MRF MODEL OF A BINARY INVERTER

Consider the MRF design of a binary NOT gate, using the technique developed in [10, 100] (Table 4.1).

4.2.1 MARGINALIZATION

Marginalization with respect to v_2 describes the probabilistic behavior of the output v_2 as follows (Fig. 4.2a):

$$p(v_2) = \frac{\exp(\frac{v_2}{kT}) + \exp(\frac{1-v_2}{kT})}{2(1 + \exp(\frac{1}{kT}))} \tag{4.4}$$

It is assumed that both output values, 0 and 1, are equally likely, and we observe two symmetrical peaks. This MRF model is implemented by a multiple feedback circuit, which computes the maximum likelihood of the compatibility function in arithmetic sum-of-product form, $\overline{x}_1 x_2 + x_1 \overline{x}_2$ (Fig. 4.2b). Note that the NOT gate can be represented in other integer-valued forms; for example, in Walsh form: $E(v) = {}^1/_4 \times [2 - 2(-1)^{v_1 + v_2}]$.

Table 4.1: Design of a noise-tolerant binary inverter based on the MRF model.

Gate	Graph model	Compatibility truth table	Energy function $E(v) = A(v)$
$x \vartriangleright\!\!\circ f$ $f = \overline{x}$	v_1 x — f v_2	v_1 v_2 U 0 0 0 0 1 1 1 0 1 1 1 0	$E(v) = v_1 + v_2 - 2v_1 v_2$

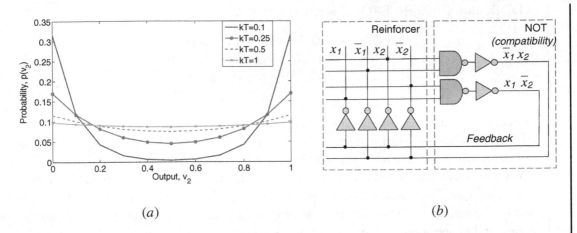

(a)   (b)

Figure 4.2: The MRF model of a binary inverter: (a) Gibbs probability distribution of the output [100] and (b) its implementation using a multiple feedback circuit [10].

4.2.2   COMPATIBILITY TRUTH TABLE

A compatibility truth table is defined by a truth table with $r + 2$ columns: $v_1, v_2, \ldots, v_r, v_{r+1}$, and a column $\mathbf{U}$, called the compatibility truth vector. The column element $u_i = 1$ if it corresponds to the desired assignment $v_1, v_2, \ldots, v_r, v_{r+1}$, and $u_i = 0$, otherwise. In order to embed a logic function in the MRF model, the compatibility truth vector $\mathbf{U}$ must be described only in integer-valued forms, such as arithmetic SOP, arithmetic polynomial, Walsh form, or Haar form [154].

4.2.3   FEEDBACK

In the MRF model, the compatibility function corresponds to a real-valued function, called the energy function, $E(x)$. It is defined as the arithmetic sum of the valid minterms, also called a pseudo-arithmetic polynomial [154]. For example, the compatibility function for NOT and EXOR (Fig. 4.1) is $U_{NOT} = x_1\bar{x}_2 + \bar{x}_1x_2$ and $U_{EXOR} = \bar{x}_1\bar{x}_2\bar{x}_3 + \bar{x}_1x_2x_3 + x_1\bar{x}_2x_3 + x_1x_2\bar{x}_3$, respectively. In general, form, the compatibility function U of an n-input function $f = x_{n+1}$ of n variables $x_1, .., x_n$ can be written as follows:

$$
\begin{aligned}
U &= u_{0...00}x_1^0...x_n^0x_{n+1}^0 + u_{0...01}x_1^0...x_n^0x_{n+1}^1 + \cdots + u_{1...11}x_1^1...x_n^1x_{n+1}^1 \\
&= \texttt{SumMaxterm} \times x_{n+1}^0 + \texttt{SumMinterm} \times x_{n+1}^1
\end{aligned}
$$

where $u_{0...00}, ...u_{1...11}$ are the components of the truth table of U, $x^0 = \bar{x}$ and $x^1 = x$, SumMaxterm and SumMinterm. For example, for the NOT function, $U_{NOT} = x_1\bar{x}_2 + \bar{x}_1x_2 = x_1\bar{f} + \bar{x}_1f$ (Fig.

4.1a). By analogy, for the EXOR function (Fig. 4.1b),

$$
\begin{aligned}
U_{EXOR} &= (\overline{x}_1\overline{x}_2 + x_1x_2)\overline{x}_3 + (\overline{x}_1x_2 + x_1\overline{x}_2)x_3 \\
&= (\overline{x}_1\overline{x}_2 \vee x_1x_2)\overline{x}_3 + (\overline{x}_1x_2 \vee x_1\overline{x}_2)x_3 \\
&= \overline{(x_1 \oplus x_2)}\,\overline{f} + (x_1 \oplus x_2)f
\end{aligned}
$$

Therefore, the compatibility function preserves the states of the MRF via this two-fold representation: a combination of maxterms (the minterms, corresponding to the values 0 of the function f), which are true when the function f is false (hence, the product of this sum and $\overline{f}$), and the minterms, which are true when the function f is true. This corresponds to an implementation of a circuit with feedback: the variable-dependent implementation of the function and its complement, both reinforced via feedback from the previous state of the system, preserved due to delay on the line (envisioned in [100] as a chain of inverters). Note that both the previous and current states correspond to one combination of the inputs, $x_1, ..., x_n$, and the output $f = x_{n+1}$.

If there is a fault on a line, such that the system "deviates" from the correct state, the reinforcement mechanism corrects the invalid state by forcing the correct state. Note that reinforcement in the form of positive feedback is ensured from the action results; that is, the more rewarding the task, the greater the probability is that it will be selected again. Such reinforcement is a well-known computing paradigm in adaptive computing and artificial intelligence [40].

4.3   MRF MODEL IMPLEMENTATION USING CYCLIC BDDS

Introduced in the 90s, and widely used in formal digital design and verification, decision diagrams have been recently implemented on emerging wrap-gate nanowire networks [58]. A decision diagram is a data structure for representing and manipulating Boolean functions. For the realization of the MRF model, binary decision diagram (BDD) techniques were proposed. The efficiency of BDD manipulation operations is determined by the number of nodes in the resulting BDD. A so-called *cyclic BDD*, with feedback, was introduced in [158]. It is suited for probabilistic computing based on the MRF model with reinforcer [100].

4.3.1   MEASURES FOR BDTS AND BDDS

BDT design.   An n-level BDT represents a logic function of n variables. A BDT is a rooted directed graph with two types of vertices: non-terminal nodes (on levels 1 to n of the tree), and terminal nodes, corresponding to the function values. Each node of a BDT represents a Shannon expansion of the implemented function with respect to one of its variables: $f = \overline{x}_i f_0 \vee x_i f_1$, where $f_0 = f_{x_i=0}$ and $f_1 = f_{x_i=1}$ correspond to the left and right outgoing branches, respectively.

The procedure for constructing a complete binary decision tree from the truth table of a Boolean function is the following.

Step 1. Derive the truth table of the Boolean function.
Step 2. Derive the first node of the decision tree as follows:

(*a*) Choose a variable to assign to the root of the tree, x_i.

(*b*) Find the Shannon expansion $f = \overline{x}_i f_{x_i=0} \vee x_i f_{x_i=1}$.

(*d*) Assign $\overline{x}_i$ to the left outgoing branch and x_i to the right one.

Step 3. Choose another variable, x_j. Construct two nodes connected to the branches of the root node: $f_{x_i=0}$ is the input of the left node and $f_{x_i=1}$ is the input of the right node. Find the Shannon expansion for both functions:

$$f_{x_i=0} \;=\; \overline{x}_j f|_{x_i=0 \atop |x_j=0} \vee x_j f|_{x_i=0 \atop |x_j=1} \quad \text{and} \quad f_{x_i=1} = \overline{x}_j f|_{x_i=1 \atop |x_j=0} \vee x_j f|_{x_i=1 \atop |x_j=1}$$

Step 4. Repeat Step 3 $(n-2)$ times, performing Shannon expansion of the factors with respect to the remaining $(n-2)$ variables.

The order of the variables can be fixed, or chosen using minimization criteria (for further reduction of the tree to a binary decision diagram).

BDD design. A BDD is an optimized BDT; that is, a BDD is a rooted directed graph, obtained from a BDT by reduction rules [154].

Implementation. BDT and BDD are representations of a discrete function by means of Shannon expansion. BDTs and BDDs are easily mapped to technology, because the circuit layout is directly determined by the shape of the BDD, and each node corresponds to a 1-to-2 demultiplexer (DEMUX) or a 2-to-1 multiplexer (MUX) (Fig. 4.3).

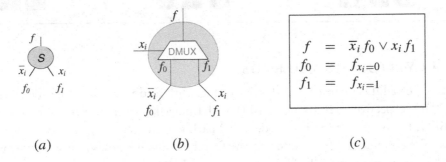

(*a*)   (*b*)   (*c*)

Figure 4.3: The node of a BDT and BDD (a), 1-to-2 DEMUX implementation (b), and formal description as Shannon expansion (c).

The terminal nodes of the BDD store the truth-table, and the output of the function is evaluated by assigning the "select" inputs of the MUX. The first design is called "top-to-bottom," while the second is referred to as "bottom-to-top." For example, decision trees and diagrams for two-input Boolean functions are given in Table 4.2.

An example of a BDD-based nanoscale structure is a wrap-gate nanowire network, in which a single or few electrons are injected into the root node, and, correspondingly, registered at one of the terminal nodes. These nanowire structures for implementing logic functions have been fabricated at the Research Center for Integrated Quantum Electronics at Hokkaido University [58].

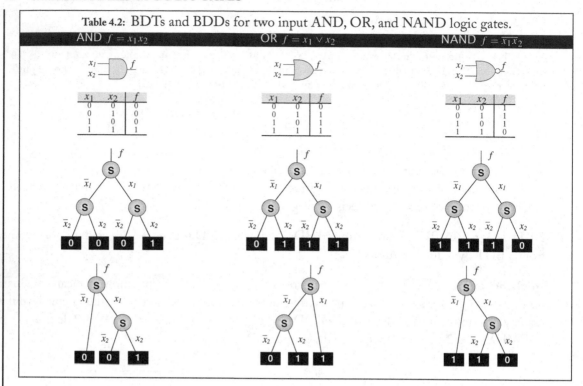

Table 4.2: BDTs and BDDs for two input AND, OR, and NAND logic gates.

4.3.2 CYCLIC BDTS AND BDDS

Any BDT or BDD can be extended to its analogs with feedback. They are called a *cyclic BDT* and *cyclic BDD*, respectively, [153, 154]. A BDT with feedback is called a *cyclic BDT*. To realize feedback in a BDT, the output of the root node is fed back with respect to a given variable to the "select" inputs of the nodes, using a reinforcement function, which can also be implemented by a BDT (Fig. 4.4). A reduced cyclic BDT is called a *cyclic BDD*. Consider, for example, a NOT gate. Its compatibility function of two variables is $U(x_1, x_2) = x_1 \oplus x_2$, where x_1 is the input, and $x_2 = \overline{x_1}$ is the output of the gate. The corresponding BDT, is shown in Fig. 4.4. Each node S of the BDT implements a Shannon expansion: $f = \overline{x_i} f_0 \vee x_i f_1$, where $f_0 = f_{x_i=0}$ and $f_1 = f_{x_i=1}$ are the left and right outgoing branches, respectively. The second BDT implements the reinforcement function, $R(x_2, U) = \overline{x_2 \oplus U}$.

4.4 SIMULATION

This section discusses the results of simulation of the proposed cyclic BDT and BDD models, using SPICE and the 16-nm Berkeley CMOS technology model (http://ptm.asu.edu/), with the gate driving voltage being $V_{DD} = 0.3\ V$ and the temperature at $100°C$ (these parameters were used

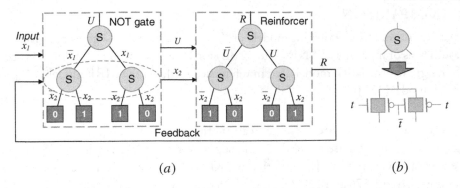

(a) *(b)*

Figure 4.4: (a) The cyclic BDT implementing the MRF model of a NOT gate, consists of two connected BDTs: the left BDT implements the compatibility function $U(x_1, x_2) = x_1 \oplus x_2$; the right BDD corresponds to the reinforcer function $R(x_2, U) = \overline{x_2 \oplus U}$; (b) Node of the cyclic BDT performs a Shannon expansion with respect to a variable; the equivalent circuit of a BDD node is a single DEMUX device.

in [100, 145]). CMOS operates in the below-threshold region, where the threshold voltages for the NMOS and PMOS are 0.4797 V and -0.4312 V, respectively.

4.4.1 NOT GATE CMOS IMPLEMENTATION USING A CYCLIC BDD.

A cyclic BDD implementation of a noise-tolerant NOT gate is given in Fig. 4.5. Similar results are derived for two-input elementary logic gates, such as NAND, NOR, AND, OR, and EXOR.

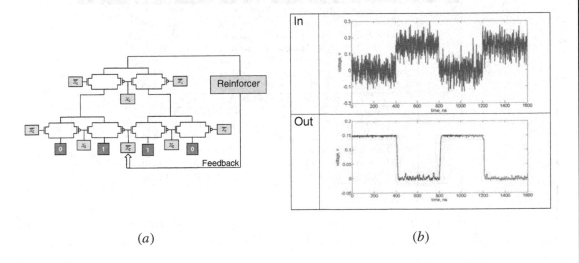

(a) *(b)*

Figure 4.5: (a) CMOS implementation of a cyclic BDT for a NOT logic gate; (b) Fragment sample of simulation results (noisy input and noise-tolerant output) at the subthreshold supply voltage (b).

4.4.2   COMPARISON

A comparison of four noise-tolerant models of the NOT and NAND gates are given in Table 4.3 for various levels of input noise (SNR =3 to 12 dB) for 16-nm CMOS technology: (*a*) conventional CMOS design, (*b*) the MRF model used in reference [100], (*c*) the MRF model [145], and (*d*) the cyclic BDT model [158].

Cyclic BDD for the NOT gate.   Let us evaluate the performance of a cyclic BDD network for a NOT gate, by modeling noise of subthreshold supply voltage while varying the SNR from 3 to 12 dB. The simulation using SPICE shows that the KLD of the cyclic BDDs ranges between 0.651 to 0.096, as opposed to 1.576 and 1.373 for e CMOS gate-level networks. The MRF model, proposed in [100], outperforms the proposed model, with the KLD varying from 0.432 to 0.012, respectively.

Cyclic BDD for the NAND gate.   The MRF model of a NAND gate, proposed in [145], is favorable in terms of KLD. However, cyclic BDDs outperform both the conventional CMOS gate-level circuits and the MRF model [100]. In terms of BER, the cyclic BDDs outperform the CMOS and MRF model [100] for lower SNR (3 to 5 dB). However, for higher SNR, all models perform comparably (Table 4.3).

Table 4.3: Comparison of noise-tolerant NOT and NAND gate MRF models, measured in KLD (BER) for various levels of SNR at 16-nm CMOS technology.

SNR	Conventional	MRF [ ]	MRF [ ]	Cyclic BDD [ ]
		Comparison for NOT gate		
3	1.576 (0.115)	0.432 (0.045)	no data	0.651 (0.059)
5	1.571 (0.062)	0.231 (0.017)	no data	0.455 (0.042)
7	1.569 (0.03)	0.116 (0.005)	no data	0.417 (0.010)
9	1.459 (0.009)	0.034 (0.005)	no data	0.298 (0.006)
10	1.455 (0.005)	0.029 (0.005)	no data	0.182 (0.005)
12	1.373 (0.002)	0.012 (0.004)	no data	0.096 (0.005)
		Comparison for NAND gate		
3	2.214 (0.103)	1.271(0.028)	0.862 (0.046)	1.246 (0.036)
5	2.171 (0.049)	0.969 (0.016)	0.508 (0.018)	0.913 (0.017)
7	1.988 (0.016)	0.723 (0.007)	0.176 (0.008)	0.566 (0.012)
9	1.922 (0.007)	0.409 (0.007)	0.032 (0.005)	0.203 (0.009)
10	1.604 (0.005)	0.14 (0.005)	0.025 (0.006)	0.124 (0.007)
12	1.400 (0.003)	0.096 (0.004)	0.007 (0.005)	0.103 (0.005)

Comparison of the performance of various two-input logic gate models based on cyclic BDDs is given in Table 4.4 and Fig. 4.6. The KLD and BER metrics are used for evaluation.

The worst-case scenario.   In the worst-case scenario, when $SNR = 3$ dB, the vulnerabilities of the proposed cyclic BDDs, starting with the most vulnerable gates, are as follows:

```
BER metric:    EXOR   OR   AND   NOT   NOR    NAND
KLD metric:    EXOR   OR   AND   NOR   NAND   NOT
```

The best noise tolerance is demonstrated by the two-input logic NAND, NOR, and NOT gate models. There is no difference in terms of the BER metric, for all gates, when $SNR = 9$ to 12 dB. However, in terms of the KLD metric, NAND and NOT gates demonstrate better noise tolerance. Compared with the standard CMOS implementation, the noise-tolerance of the cyclic BDDs, measured in terms of the BER metric, is two to three times better, in the worst-case scenario where $SNR = 3$. For example, the cyclic BDD model of the NAND gate is 2.9 times more noise tolerant in terms of BER (0.036 for BDD versus 0.103 for CMOS), if $SNR = 3$dB.

Table 4.4: Noise tolerance of cyclic BDD models of two-input logic gates, measured in terms of KLD and BER for various levels of SNR, simulated for 16-nm CMOS technology.

SNR	AND		OR		NOR		EXOR		NAND		NOT	
	KLD	BER	KLD	BER	KLD	BER	KLD	BER	KLD	BER	KLD	BER
3	2.05	0.124	2.73	0.189	1.68	0.035	3.38	0.291	1.25	0.036	0.65	0.059
5	1.83	0.025	2.47	0.136	1.48	0.028	2.40	0.167	0.91	0.017	0.46	0.042
7	1.60	0.010	1.95	0.066	1.28	0.009	1.81	0.072	0.57	0.012	0.42	0.010
9	1.47	0.013	1.59	0.020	1.16	0.007	1.50	0.021	0.20	0.009	0.30	0.006
10	1.38	0.012	1.53	0.013	1.14	0.006	1.35	0.021	0.12	0.007	0.18	0.005
12	1.28	0.009	1.12	0.007	0.94	0.006	0.93	0.019	0.10	0.005	0.10	0.005

4.5   NOISE-TOLERANT TWO-BIT ADDER

This section considers the efficiency of implementing shared BDDs using the MRF model. Shared BDDs [83] are used to represent multi-output functions. A two-bit adder is a three-output function. In order to form its MRF model on BDD, its shared BDD is extended here by adding feedbacks, thus forming a *cyclic shared BDD*. The area, power, and delay of the CMOS models (traditional and MRF-based from [100]) of a 2-bit adder are compared against the CMOS MRF model, implemented by a shared cyclic BDDs. The results of experiments with large benchmarks from the MCNC91 set show that shared cyclic BDDs demonstrate lesser total area and power consumption, while providing superior noise immunity.

4.5.1   SHARED BDDS

The two products in Equation 4.5 are implemented by two separate BDDs, and the reinforcer is implemented by a separate BDD, thus forming a cyclic BDD structure (Fig. 4.7). Switches are also needed to ensure the external update of the system, corresponding to the state update if another combination of input and output is evaluated. Switches resolve the problem of contention of the feedback values with the initial circuit inputs (such contention exists in the PLA-based structure

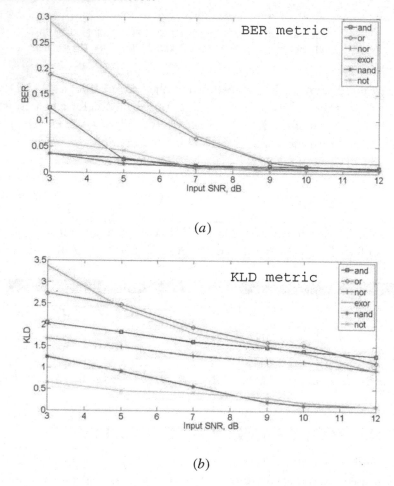

(a)

(b)

Figure 4.6: (a) Comparison of noise-tolerance of cyclic BDD models of two-input logic gates, measured in terms of the BER metric; (b) KLD metric with respect to the input SNR for the 16-nm CMOS technology.

proposed in [100]). Therefore, the MRF model of a logic function is implemented by a BDD for $U_{NOT}(x)$, consisting of two BDDs (for the maxterms and minterms of its compatibility function, respectively), a BDD for the reinforcer $R_{NOT}(x)$, and switches.

Shared BDDs were introduced in [83]. These BDDs are used for multiple-output circuits. For example, shared BDDs were used in [38] for the design of a two-bit ALU on a wrap-gate nanowire BDD network. Fig. 4.8 shows the shared BDD of a two-bit adder.

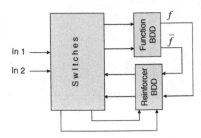

Figure 4.7: Noise-immune implementation of the MRF model on cyclic BDDs.

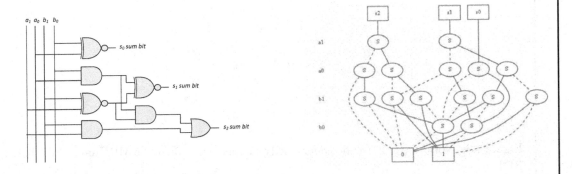

Figure 4.8: A two-bit adder (left) and its implementation using a shared BDD (right).

4.5.2 SHARED CYCLIC BDD

A shared cyclic BDD for a two-bit adder is designed by adding a block of reinforcers to each output (Fig. 4.8). Note that the switches are not shown, but assumed, in this figure.

4.6 EXPERIMENTAL STUDY

The MRF model of a multi-output switching function based on a shared cyclic BDD structure was simulated using SPICE and the 16-nm Berkeley CMOS technology model (http://ptm.asu.edu/), with the gate driving voltage being $V_{DD} = 0.3\ V$ and the operating temperature being room temperature, $27°C$. The transistors operated in the below-threshold region, where the threshold voltages for the NMOS and PMOS were $0.4797\ V$ and $-0.4312\ V$, respectively. For shared BDD design, the CUDD package was used (http://vlsi.colorado.edu/fabio/CUDD/). The experiment was designed to study the following cases:

1. Behavior of our circuit in presence of noise, as compared with a conventional CMOS design of a two-bit adder. Previous studies were limited to logic gates (two-input, one-output devices);

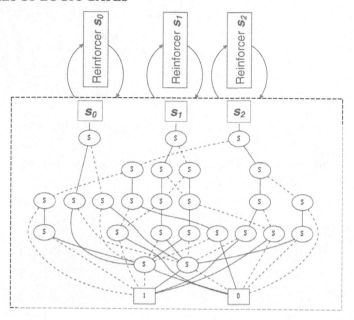

Figure 4.9: Noise-tolerant two-bit adder based on implementation of the MRF model by a shared cyclic BDD.

here, four-input, three-output circuits are considered, and significant gains in noise tolerance are expected.

2. The power dissipation, delay, and number of transistors of BDD-based designs compared with CMOS designs.

3. Feasibility of the approach based on MRF model, using shared cyclic BDDs for more complicated circuits.

4.6.1   COMPARISONS OF TWO-BIT ADDERS

To evaluate the performance of various models, SNR, the BER, and the KLD metrics were used. In Table 4.5, we show a comparison between a conventional CMOS design for a two-bit adder, and a noise-tolerant design based on an MRF model implemented by a shared cyclic BDD using 16 nm CMOS technology. In this experiment, the input noise was varied from SNR = 3 dB to SNR = 12 dB.

As expected, the MRF model of a 2-bit adder implemented by shared cyclic BDD outperforms the standard CMOS design. In particular, the improvement in BER is around 50% to 80% and for the KLD values, the improvement is around 70% to 90%. This noise-tolerance effect is comparable with the results reported for logic gates and circuits in [100, 145, 158].

Table 4.5: Comparison of conventional 2-bit adder and MRF model of 2-bit adder (implemented using a shared cyclic BDD) in terms of KLDs and BERs for various levels of SNRs simulated with 16 nm CMOS technology.

SNR	Conventional CMOS 2-bit adder						MRF model of 2-bit adder					
	Output s_2		Output s_1		Output s_0		Output s_2		Output s_1		Output s_0	
	BER	KLD	BER	KLD	BER	KLD	BER	KLD	BER	KLD	BER	KLD
3	0.072	0.752	0.440	2.475	0.233	2.350	0.080	0.220	0.146	1.488	0.147	1.602
5	0.054	0.591	0.254	2.128	0.150	2.245	0.024	0.079	0.047	1.227	0.057	0.879
7	0.043	0.375	0.103	1.795	0.098	1.955	0.012	0.029	0.022	0.666	0.022	0.471
9	0.042	0.359	0.078	1.663	0.097	1.607	0.009	0.009	0.015	0.351	0.012	0.430
10	0.042	0.328	0.068	1.484	0.093	1.501	0.008	0.007	0.017	0.289	0.011	0.271
12	0.036	0.185	0.059	1.339	0.091	1.368	0.005	0.007	0.016	0.210	0.011	0.228

Fig. 4.10 illustrates the high noise-immunity of the shared cyclic BDD implementation of the MRF model, compared with a conventional design, given a fixed SNR value of 7 db for the input signal.

4.6.2 AREA, POWER, AND DELAY

In Table 4.6, the power dissipation, delay, and the number of transistors are compared for the standard CMOS design and the MRF model implementations reported in [100, 145, 158]. The operating voltage for the CMOS based design is $0.7V$ as indicated by the ITRS roadmap [49]. The BDD based noise-tolerant models were operated in the subthreshold region with the transistor driving voltage being $0.3V$.

Table 4.6: Performance comparison of the conventional CMOS design of a 2-bit adder, and its noise-tolerant designs using an MRF model reported in [100, 145, 158], and the proposed shared cyclic BDD implementation.

Parameter	Conventional	MRF [ ]	MRF [ ]	MRF [ ]	MRF BDD
Supply voltage (v_{dd})	0.7	0.3	0.3	0.3	0.3
No. of transistors	60	420	196	176	152
Power dissipation (nW)	0.151	0.042	0.034	0.039	0.034
Output delay avg. (nS)	3	16.4	11.1	21.2	18.3

4.6.3 SIMULATION RESULTS

Shared BDDs are preferable for large circuits [83, 154]. To study the effectiveness of shared cyclic BDDs in the implementation of MRF models, MCNC'91 benchmarks were used (Table 4.7), and the following settings were adopted: the conventional CMOS designs were simulated for the $V_{dd} = 0.7V$ and MRF designs for $V_{dd} = 0.3V$.

Table 4.7 shows MCNC'91 benchmark circuit specifications in terms of number of transistors and power dissipation characteristics, for various numbers of inputs and outputs and design methodologies.

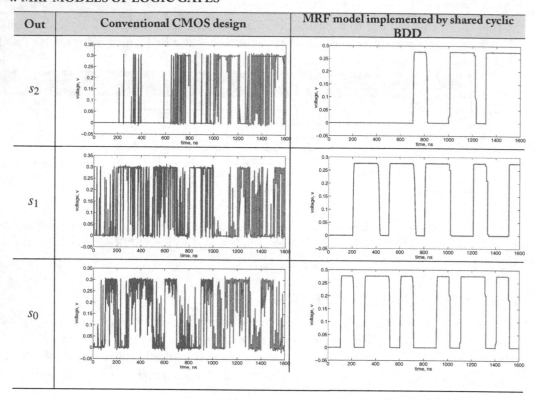

Figure 4.10: Noise output signals s_0, s_1, and s_2 of the 2-bit adder for an input SNR value of 7 db for conventional CMOS design (left) and the same noise-free signals in a CMOS MRF model implemented by shared cyclic BDD (right).

Table 4.7: Comparison of number of transistors and power dissipation required for conventional CMOS implementation, MRF model [100], and proposed design for circuits from MCNC'91 benchmark base.

MCNC'91 benchmarks			CMOS design $V_{dd} = 0.7V$		MRF design [ ] $V_{dd} = 0.3V$		Proposed design $V_{dd} = 0.3V$	
Circuits	Input	Output	# tran	Power (nW)	# tran	Power (nW)	# tran	Power (nW)
5xp1	7	10	568	1.431	2756	0.487	608	0.138
alu4	14	8	6928	17.458	33416	5.915	7166	1.626
con1	7	2	78	0.196	356	0.063	120	0.027
ex5	8	63	5448	13.728	25964	4.596	6108	1.386
cordic	23	2	604	1.522	2612	0.462	720	0.163
misex1	8	7	356	0.897	1700	0.300	386	0.087
rd53	5	3	236	0.594	1012	0.179	284	0.064
squar5	5	8	346	0.871	1532	0.271	356	0.080

4.7 TERNARY INVERTER

A ternary, or three-valued, inverter is a basic element for ternary storage devices. Our interest in designing a noise-tolerant ternary inverter is motivated by new technological advances. Ternary logic gates are known to have advantages over their binary counterparts, such as increased capacity of transmitted information, decreased number of connections, as well as favorable power dissipation, delay, and circuit complexity characteristics. A binary NOT gate is a basic element for memories; in particular, the CMOS Static Random-Access Memory (SRAM) cells. The SRAM utilizes static latches as the storage cells, which should be sized as small as possible in order to achieve high memory density. A typical SRAM cell consists of a cross-coupled CMOS inverter pair (six PMOS transistors or four NMOS transistors). Keeping the static power dissipation per cell as low as possible is a priority in SRAM cell design.

CMOS ternary logic is discussed in [147]. The ternary NOT gate for building SRAM devices was studied in [55]. In an ideal inverter, both the noise margins and the threshold voltage are equal to $V_{dd}/2$, where V_{dd} is the power supply voltage [119].

4.7.1 NOISE-TOLERANT TERNARY INVERTER

Consider a ternary inverter whose input v_1, which output is calculated as follows: $v_2 = 2 - v_1$. Let us design the compatibility truth table (Table 4.8), and using vector $\mathbf{U}$, find the compatibility function in arithmetic form: $f(v) = \frac{1}{2}(-v_2 + v_2^2 - v_1 + 11v_1v_2 - 6v_2^2v_1 + v_1^2 - 6v_1^2v_2 + 3v_1^2v_2^2)$ [125]. Embedding this function into the model (Equation 4.1) corresponds to the equivalence $f(v) = E(v)$. After substitution of $E(v)$ into the model (4.1), we obtain the Gibbs joint probability distribution: $p(v_1, v_2)$. Summing over all possible values of the variable v_1 eliminates this variable, leaving the marginal probability distribution of the output v_2:

$$
\begin{aligned}
p(v_2) &= \frac{1}{Z_1} \sum_{v_1 \in 0,1,2} \exp\left(-\frac{E(v)}{kT}\right) \\
&= \frac{1}{Z_1}\left[\exp\left(\frac{-2v_2 + 2v_2^2}{4kT}\right) + 2\exp\left(\frac{8v_2 - 4v_2^2}{4kT}\right) + 2\exp\left(\frac{4 - 6v_2 + 2v_2^2}{4kT}\right)\right]
\end{aligned}
$$

A family of distributions $p(v_2)$ for various parameters kT is given in Table 4.8. One can observe that the output values are grouped around logic values 0, 1, and 2 with equal probability (three picks); this is because it is assumed that the input is equally likely to be 0, 1, or 2. Therefore, the Gibbs distribution provides an acceptable probabilistic approximation of uncertainty, caused by noise.

4.7.2 TERNARY CMOS NOT AND MIN-NOT GATE

The results of simulation of the proposed MRF-based ternary inverter, and of a CMOS-based one, are given in [55].

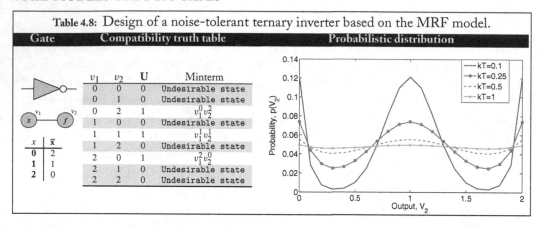

Table 4.8: Design of a noise-tolerant ternary inverter based on the MRF model.

Gate	Compatibility truth table				Probabilistic distribution

Compatibility truth table:

v_1	v_2	U	Minterm
0	0	0	Undesirable state
0	1	0	Undesirable state
0	2	1	$v_1^0 v_2^2$
1	0	0	Undesirable state
1	1	1	$v_1^1 v_2^1$
1	2	0	Undesirable state
2	0	1	$v_1^2 v_2^0$
2	1	0	Undesirable state
2	2	0	Undesirable state

x	$\bar{x}$
0	2
1	1
2	0

Ternary inverter. In the CMOS model, a high-resistance transmission gate is placed between a low-resistance inverter and 0.3 V supply, in order to produce a middle voltage level (Fig. 4.11a). The threshold voltage of transistors $P2$ and $N2$ is half of the supply voltage. Note that both $P2$ and $N2$ are in the subthreshold region for the middle state "1." A typical problem of this CMOS implementation (and other known ternary gates, for example, [147]), is that the noise margins of the voltage transfer characteristic decrease. This is a well-known effect, observed in many multi-valued circuit realizations: that for fixed values of the highest and lowest voltages, the noise-tolerance of a circuit with more logic levels is worse, compared to its binary analogue.

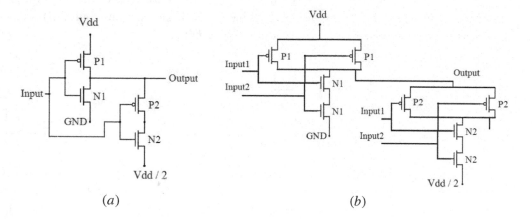

Figure 4.11: (a) Conventional CMOS ternary inverter [55] and (b) Conventional CMOS two-input ternary MIN-NOT gate.

Embedding the ternary inverter logic function. The arithmetic sum of minterms form of the compatibility function of a ternary inverter is derived from its compatibility truth table (in which there are three valid state combinations (02,11,20), assigned the value 1) as follows

$$\mathbf{U} = \frac{1}{4}(v_1^0 v_2^2 + v_1^1 v_2^1 + v_1^2 v_2^0) \tag{4.5}$$

where the literal operation is defined as $y^x = 2$ if $x = y$ and $y^x = 0$ otherwise, $x, y \in \{0, 1, 2\}$.

To implement an MRF model that embodies the above sum form (Equation 4.5), and uses the maximum likelihood principle and multiple feedback, we need a conventional ternary inverter and a conventional two-input ternary MIN-NOT gate. A simple ternary inverter is shown in Fig. 4.11 [55]. The inverter has one input and one output that can take three different logic values, 0, 1 and 2. The logic value 0 is represented by the voltage $0V$, and logic levels 1 and 2 correspond to $0.15V$ and $0.3V$, respectively.

Noise-tolerant ternary MIN-NOT gate. A ternary two-input NAND or MIN-NOT gate, $\overline{\text{MIN}}(x_1, x_2)$, is shown in Fig. 4.11b. In this circuit, $P1$ and $P2$ are the p-channel enhancement transistors, whereas $N1$ and $N2$ are n-channel enhancement transistors. The threshold voltage V_T and supply voltage, V_{dd} are specified by the following conditions: $0 < |V_T| < |V_{dd}|$ for $P2$ and $N2$ and $|V_{dd}| < |V_T| < |2V_{dd}|$ for $P1$ and $N1$.

Noise-tolerant ternary inverter. The circuit for a noise-tolerant ternary inverter is shown in Fig. 4.12. Different state combinations of the storage nodes ensures the propagation of conditional probability within the circuit. Assigning the highest probability to a valid state combination is ensured by the reinforcement feedback. The feedback forces the circuit to return to state combinations for which the clique energy function is true, thus making the system tolerant to incorrect output due to noise. For example, suppose the nodes, v_1 and v_2, take the logic values 0 and 2, respectively. Via the feedback from cyclic inverters [125], the inputs of the MRF based circuit assume any of the possible states. The first ternary MIN-NOT gate stays inactive and feeds logic level 0 to the corresponding input nodes. The second and third MIN-NOT gate outputs are logic 1 and logic 2, which are consistent with the storage node values. For the other valid state combinations between the two storage nodes, the stability of the network can be ensured accordingly.

The experimental study involves measurement of the noise-tolerance characteristics of the proposed MRF-based model of a ternary inverter by various metrics, and also comparison of the latter against the conventional CMOS design of the inverter.

Simulation using 16-nm CMOS technology. The proposed noise tolerant model was simulated using SPICE and the 16-nm Berkeley CMOS technology model (http://ptm.asu.edu/), with the gate driving voltage being $V_{DD} = 0.3$ V at room temperature, $27°C$. CMOS models with two different threshold voltages were used for the ternary experiments. The CMOSs denoted as P1 and N1 are the regular PMOS and NMOS models with threshold voltages 0.4797 V and -0.4312 V respectively. The P2 and N2 models represent the PMOS and NMOS with threshold voltages being of -0.17 V and 0.17 V, respectively. When the supply voltage is lower than the threshold voltage

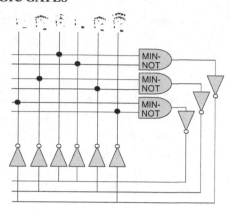

Figure 4.12: Noise-tolerant ternary inverter based on MRF model

of the transistor, the following noise effects arise: noise margin reduction caused by thermal noise, electromagnetic coupling, hot-electron effects, and threshold variations.

4.7.3   COMPARISON WITH CONVENTIONAL CMOS DESIGN

Table 4.9 contains the results of the comparison of a conventional CMOS design for a ternary inverter [55] and the proposed noise-tolerant design. The latter is based on the MRF model, implemented using 16 nm CMOS technology. In this experiment, the input SNR was varied between 9 dB and 18 dB.

Table 4.9: Comparison of the ternary conventional and noise-tolerant inverters.

SNR	Conventional CMOS design [ ]		MRF-based design	
	BER	KLD	BER	KLD
9	0.2484	3.9198	0.0861	1.7438
10	0.2301	3.8697	0.0364	1.4980
12	0.2103	3.7170	0.0191	1.2048
14	0.1824	3.6329	0.0164	1.0628
16	0.1726	3.5169	0.0102	0.7508
18	0.1493	3.3761	0.0100	0.5543

As expected, the MRF model of a ternary inverter outperforms the conventional CMOS design. In particular, the improvement in BER is 60% - 94%, and for the KLD values, the improvement is 55% - 83%. Fig. 4.13 illustrates strong noise-immunity of the proposed implementation of the MRF model, compared to a conventional design, given the fixed SNR value of 10 dB for the input signal.

We measured a distribution of the output voltage probability for the proposed noise-tolerant ternary inverter, given an input signal SNR of 9 dB and 16 dB (Fig. 4.14). One can observe that

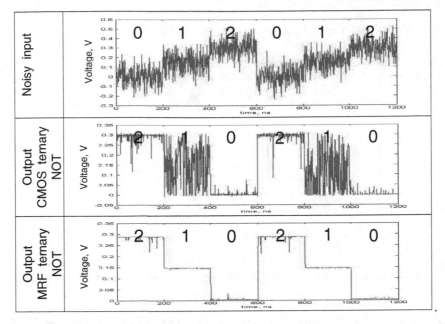

Figure 4.13: The output of a conventional ternary CMOS NOT gate, and the noise-tolerant MRF-based model, given the noisy input signals

random variations of voltage are concentrated around $0V$, $0.15V$, and $0.3V$, corresponding to the logical values "0," "1," and "2," respectively. Hence, both the theoretical MRF model with Gibbs distribution (Table 4.8), and the measured probability distribution at the output of CMOS circuit, which operates accordingly to the maximum likelihood principle, exhibit a very close resemblance.

4.7.4 AREA, POWER, AND DELAY

In Table 4.10, power dissipation, delay, and number of transistors are shown for both CMOS and MRF models of the inverter. The operating voltage for the CMOS design is $0.7V$, as indicated by the ITRS roadmap [49]. The noise-tolerant ternary inverter operates in the sub-threshold region; the transistor driving voltages are $0.3V$ and $0.15V$.

Table 4.10: Comparison of CMOS conventional and noise-tolerant ternary inverter		
Parameter	Conventional design	MRF based design
Supply voltage (V_{dd}) (V)	0.7/0.35	0.3/0.15
Number of transistors	4	56
Power dissipation (nW)	0.378	0.048
Output delay (nS)	0.3334	0.6667

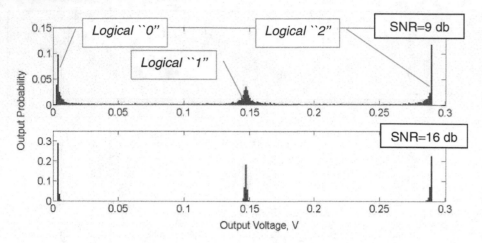

Figure 4.14: Output voltage probability distribution of a noise-tolerant ternary inverter.

Although in the MRF model the number of transistors increases compared to the conventional CMOS based design, the power dissipation does not increase, given the small driving voltages. Instead, an 87% reduction in power consumption is observed, on average. At the same time, the average delay for the MRF based ternary circuits is less, compared to binary ones with similar models.

4.8 SUMMARY AND DISCUSSION

This chapter contributes to the application of MRF models to noise-tolerant logic gate design. In these lectures we reviewed the MRF model of logic gates as originally expounded in [10], but which was perhaps explained more lucidly in [98, 99, 100]. It was theoretically justified and experimentally proved that MRF-based logic gates and networks are able to operate in the presence of intense noise. The MRF models are reasonable candidates for CMOS implementation. Specifically, the following results of ours contribute to the area of noise-tolerant design of logic gates and networks:

1. The multi-feedback reinforcer [100] is an efficient implementation of the MRF computing model. We introduce an extension of this design technique based on BDDs.

2. The proposed feedback schemes and MRF models of logic circuits are compatible with computer aided design, physical layout, and implementation of ICs on nanoscale micro- and nano-electronic devices. For example, quantum-effect resonant-tunneling devices, photonic devices, wrap-gate nanowire BDD circuits [38], resistor-logic demultiplexers [67], as well as other solutions, can be examined. The proposed design is suitable for noise- and fault-tolerant BDD-centric networks, because robust cyclic BBDs may correct and accommodate soft tech-

nological and operational errors. Another extension of the MRF models and cyclic BDDs is their applicability and capacity to be generalized from 2D to the 3D nanoscale structures [75].

3. MRF models for multivalued logic circuits can be useful for predictive technologies. We demonstrate the MRF model of three-valued logic gates implemented by cyclic BDDs.

4. We show that the design of noise-tolerant logic circuits is based on a conventional taxonomy using noise-tolerant logic primitives (gates). For this reason, we focus on the design of noise-tolerant gates and simple adders.

CHAPTER 5

Neuromorphic models

Murphy's principle of probabilistic computation

The probability of anything happening is in inverse ratio to its desirability

Artificial neural models are understood as biologically inspired computing networks and memory storage; they aim to exploit the properties of the brain such as robustness and fault tolerance, flexibility, processing of fuzzy, probabilistic, noisy, or inconsistent information, and massively parallel and distributed processing [82]. Neuromorphic networks are a hardware implementation of artificial neural networks; they resemble cooperative phenomena and can process probabilistic, noisy, or inconsistent information. The familiar examples of these models include the Hopfield network and its improved successor known as the Boltzmann machine.

In this chapter, we use Hopfield and Boltzmann models for the design of noise-tolerant logic gates and networks. Similarly to the MRF, the Hopfield computing paradigm is based on the concept of *energy minimization* in a stochastic system. However, locality is an inherent property of the Hopfield model, in contrast to the MRF model, in which the neighborhood must be specified by a set of cliques. A similar design principles should be used for the design of networks of logic gates, such as connections of MRF or Hopfield models of logic gates.

The Hopfield (Boltzmann) computing paradigm is based on a property of threshold cells in a distributed system called *parallel relaxation*: computing by a network of threshold cells using the random change of their states, given an objective function, until the stable states are achieved. These stable states encode the final result. In the modeling of logic gates, this result is contained in the values of the logic function. Relaxation can be interpreted as a state search in a search space of all possible states. A randomly chosen state will transform itself ultimately into one of the local minima (the nearest stable state). Even if the initial state contains inconsistencies, a network will settle into a solution that violates the fewest constraints offered by inputs.

5.1 HOPFIELD NETWORK AND BOLTZMANN MACHINE

This section is a brief introduction to the computing of logic functions using Hopfield networks and Boltzmann machines. The basic computing element in these models is the neuron cell, also called a *threshold cell*. Threshold cells were developed by McCulloch and Pitts in 1943 and are specified by integer-valued operations and thresholding ("A logical calculus of ideas immanent in nervous activity," Bulletin of Mathematical Biophysics, vol. 5, no. 115, 1943). This pioneering work ushered in the modern era of neural networks.

5.1.1 THRESHOLD GATE

A switching function f of n variables $x_1, x_2, \ldots, x_n$ is called a *threshold* function with weights $w_1, w_2, \ldots, w_n$ (real numbers) and threshold k if $f = 1$ if and only if $\sum_{i=1}^{n} w_i \times x_i \geq k$. A threshold function of two variables is implemented by a threshold cell (as shown in Figure 5.1). This threshold gate can compute various elementary switching functions of two variables. For this, appropriate control parameters must be chosen. For example, the n-input OR and AND functions have thresholds of 1 and n, respectively. That is, in this threshold gate, increasing the number of active inputs results in the output 1.

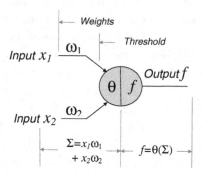

Figure 5.1: The two-input threshold cell: the input signal for $x_1, x_2 \in \{0, 1\}$ is $\Sigma = x_1 \times w_1 + x_2 \times w_2$; thresholding of Σ at the level Θ produces the output signal $f = \Theta(\Sigma)$.

An arbitrary logic function can be computed on a network of threshold cells. Table 5.1 shows a technique for computing an elementary logic function using threshold cells, where $\sum$ is an arithmetic sum. The threshold cell is very flexible for functional reconfiguration. For example, two-input AND, OR, NOR, and NAND elementary logic functions can be generated by the same threshold cell by choosing appropriate control parameters, such as the threshold θ and weights $w_i \in \{1, -1\}$ of the arithmetic sum $w_i \times x_1 + w_i \times x_2$. The output is $f = w_1 \times x_1 + w_2 \times x_2 - \theta$. Let the threshold be $\theta = -2$, and $w_1 = w_2 = -1$, then the sum of the weighted inputs is $(-1) \times x_1 + (-1) \times x_2$. Given the assignment $x_1 = 1, x_2 = 0$, the output is $(-1) \times 1 + (-1) \times 0 - (-2) = 1$. Note, that the weights take only two values, 1 and -1, which is ideal for implementation.

Table 5.1: The McCulloch–Pitts model for computing AND, OR, and NAND switching functions.

AND	OR	NAND

$f = x_1 x_2$ $\qquad\qquad$ $f = x_1 \vee x_2$ $\qquad\qquad$ $f = \overline{x_1 x_2}$

x_1	x_2	$\sum$	f
0	0	0	0
0	1	1	0
1	0	1	0
1	1	2	1

x_1	x_2	$\sum$	f
0	0	0	0
0	1	1	1
1	0	1	1
1	1	2	1

x_1	x_2	$\sum$	f
0	0	0	1
0	1	-1	1
1	0	-1	1
1	1	-2	0

5.1.2   NOISE AND MEASURE OF UNCERTAINTY

In modeling, noise, η, with mean of zero and variance σ_η^2 can be injected in a threshold cell in various ways. The uncertainty of the threshold output can be measured in terms of mutual information. In our experiments, we used the *processing noise* and *additive input noise* models of threshold cells. Both these types of noise injection are related to the states of threshold cells.

Processing noise.   *Processing noise* (also called *functional noise*) is defined as $y = w_1 x_1 + w_2 x_2 + \eta$, where the output y is a random variable with normal probability distribution and variance σ_y^2 assuming the inputs x_1 and x_2 are normally distributed random variables. The mutual information of the threshold cell with processing noise is $I(y; x_1, x_2) = (^1/_2) \log(\sigma_y^2 / \sigma_\eta^2)$ [40]. It is a measure of the uncertainty of the output y resolved by observing the inputs x_1 and x_2. Note that the ratio $\sigma_y^2 / \sigma_\eta^2$ is called the *Signal-To-Noise Ratio* (SNR). The mutual information $I(y|x_1, x_2)$ is maximized by maximizing the SNR.

Additive input noise.   An *additive input noise* (also called *connection noise* because it is injected into connections between threshold cells) is defined as $y = w_1(x_1 + \eta_1) + w_2(x_2 + \eta_2)$, where η_1 and η_2 are assumed to be independent, normally distributed random variables. Hence, it can be assumed that the noise distribution in a threshold cell is a joint (multivariate) normal distribution. The mutual information of the threshold cell with processing noise is $I(y; x_1, x_2) = (^1/_2) \log \frac{\sigma_y^2}{\sigma_\eta^2(w_1^2 + w_2^2)}$ [40]. Hence, this a measure of the uncertainty of the output y resolved by observing the inputs x_1 and x_2 in this kind of injected noise. The maximization of mutual information $I(y; x_1, x_2)$ leads to a reduction in redundancy in the output y as compared to that in the inputs x_1 and x_2.

5.1.3 NETWORK OF THRESHOLD CELLS

Threshold networks can be considered as an alternative to logic networks. In some cases, networks of threshold elements may be more efficient than logic networks. For example, the threshold network shown in Fig. 5.2a is used for the implementation of AND-OR and OR-AND logic networks on threshold networks as shown in Fig. 5.2b, c. The control parameters are taken from Table 5.1 as follows: (*a*) An AND-OR network (Fig. 5.2b) is designed from a threshold network by specifications of the weights and thresholds for an AND function; $\omega_{11} = \omega_{12} = \ldots = \omega_{43} = 1$, $\theta_1 = \theta_2 = \theta_3 = 2$, $\theta_4 = 1$. (*b*) An OR-AND network (Fig. 5.2c) is designed by analogy: $\omega_{11} = \omega_{12} = \ldots = \omega_{43} = 1$, $\theta_1 = \theta_2 = \theta_3 = 1, \theta_4 = 2$.

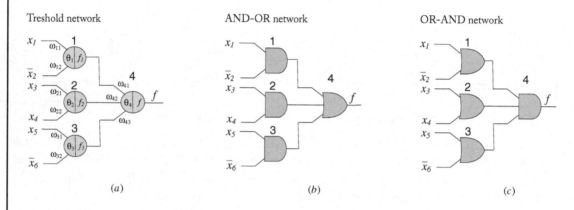

Figure 5.2: A network of threshold cells (a) and the corresponding AND-OR (b) and OR-AND (c) logic networks.

5.1.4 HOPFIELD NETWORK

In 1984, Hopfield utilized the concept of energy minimization in a stochastic system to formulate a new computation paradigm using a recurrent network of threshold elements of particular configuration, known as a Hopfield network (model) [44, 45]. The Hopfield network is a *content-addressable memory*. The primary function of the Hopfield network is to retrieve a pattern stored in memory in response to the presentation of an incomplete or noisy version of that pattern. A content-addressable memory is *error-correcting* in the sense that it can override inconsistent information in presented patterns containing partial but sufficient information.

The basic processing unit of the Hopfield network is a threshold cell or McCullock and Pitts neuron, which has two states encoded by "0" and "1." Instead of memory devices, such as NOT gates with feedback and flip-flops with two feedbacks, a threshold cell has no self-feedback. In a Hopfield network, threshold cells store information (in weights, thresholds, and particular topology) in a *distributed* (or redundant) manner. The value of a switching function, given an assignment of its binary variables, is computed through relaxation of the neuron cells in the network.

In a discrete Hopfield network, (a) the output of each threshold cell is fed back to all other threshold cells, (b) there is no self-feedback ($w_{ii} = 0$), and (c) the weights of connections are symmetric ($w_{ij} = w_{ji}$) (Fig. 5.3a). The notation of "states" is central in the description of Hopfield models: state of a threshold cell, state of a network of Hopfield models (gates), initial state, stable state, local and global state.

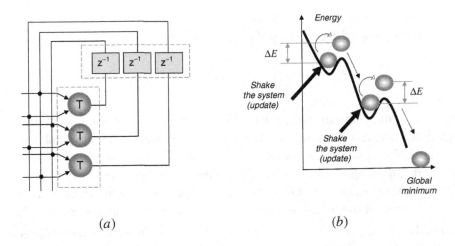

(a) $\qquad\qquad\qquad\qquad\qquad\qquad (b)$

Figure 5.3: (a) Hopfield and Boltzmann model have the same architecture, such as a 3-loop feedback network, where cells are threshold elements T and z^{-1} are delay elements; (b) In contrast to the Hopfield model, that explores local minima, the Boltzmann model avoids local minima by "shaking" the current state of threshold cell.

In the absence of noise, the Hopfield model provides for reliable computation of a specified elementary logic function. That is, the network of threshold cells can achieve global minimum for an arbitrary assignment of input variables. However, if a Hopfield model computes under noise, the correct output is computed with some probability. Thus, the results of computation are unreliable and must be improved; for example, by repeating computations. The *convergence time* in Hopfield model is defined as the number of discrete-time updates before the model converges (within a given precision for analog states). In a stable state (local or global) of the Hopfield gate, energy is at a minimum. There is no way to reach a global minimum from a local minimum without a special mechanism.

Details of the operation of the Hopfield network (the storage phase and the retrieval phase), analysis of storage capacity related to local minima, conditions for stability, and probabilistic analysis using the signal-to-noise ratio metric and conditional probability distribution of bit error can be found in [40].

5.1.5   BOLTZMANN MACHINE

The original Hopfield network uses local minima as the memory of a network. The Boltzmann machine can be viewed as an improved Hopfield model. The goal of the Boltzmann machine is to reach a global energy minimum instead of exploring the local minima like the memory in the Hopfield model.

A Boltzmann machine explores the isomorphism between a system with feedback (a Hopfield network), and probability distribution of states used in statistical physics. Let the states of a threshold cell be x_1 and x_2. Then the probability of any global state is constant and obeys the Gibbs (Boltzmann) distribution

$$\frac{p(x_1)}{p(x_2)} = \exp \frac{-(E(x_1) - E(x_2))}{T}$$

where T is a control parameter. That is, the probability ratio of any two states depends on their energy difference, $E(x_1) - E(x_2)$, and the lowest energy state is the most probable.

In a Boltzmann machine, noise is used to "shake" the gate state out of a local minimum (Fig. 5.3b). In this technique, local minima are avoided by adding some randomness to the process of energy minimization, so that when the network moves toward a local minimum, it has a chance to escape. Updating of individual cells is a stochastic process. Specifically, a cell v_k has an active state 1 with probability $p_k = 1/1 + e^{-\frac{\Delta E_k}{T}}$, where $\Delta E \in \{+, -, 0\}$ is the energy that corresponds to the state change of the cell. The output of a threshold cell is always either logical 0 or 1.

5.1.6   LOGIC FUNCTION EMBEDDING

A logic function is embedded into the Hopfield model via the energy function and the compatibility function (truth table). The local properties of graphical data structure of threshold elements are embedded in a complete undirected graph; the neighboring node interconnectivity is an inherent property of the Hopfield model. Hence, the embedding mechanism for representing particular logic properties into the Hopfield model is more simple than the MRF model.

The energy function of the Hopfield model is defined as follows:

$$E = -\frac{1}{2} \sum_{i}^{n} \sum_{j}^{n} x_i x_j C_{ij} + \sum_{i}^{n} x_i N_i,$$

where x_k is the state of cell k, C_{ij} is the connection weight between the cells i and j, N_i is the threshold value of the ith threshold cell, and n is the total number of cells, excluding the *Bias* cell. All the global minima of the energy function correspond to "True" values of the switching function, while "False" values are encoded as having an energy value greater than the global minimum. In the process of relaxation, the Hopfield networks try to achieve the global minimum, and, therefore, the correct output.

A set of fundamental two-input logic functions can be implemented as a three-node Hopfield network. The Hopfield models of elementary logic functions are shown in Table 5.2.

Table 5.2: Design taxonomy of Hopfield models of logic gates.

Gate	Energy function, E	Model	Compatibility truth table

AND

$f = x_1 x_2$

$E = -v_1 - v_2 + 2v_3 + v_1 v_2 - 2v_1 v_3 - 2v_2 v_3$

v_1	v_2	v_3	E	f
−1	−1	−1	−3	−1
−1	−1	1	9	False
−1	1	−1	−3	−1
−1	1	1	1	False
1	−1	−1	−3	−1
1	−1	1	1	False
1	1	−1	1	False
1	1	1	−3	1

OR

$f = x_1 \vee x_2$

$E = v_1 + v_2 - 2v_3 + v_1 v_2 - 2v_1 v_3 - 2v_2 v_3$

v_1	v_2	v_3	E	f
−1	−1	−1	−3	−1
−1	−1	1	1	False
−1	1	−1	1	False
−1	1	1	−3	1
1	−1	−1	1	False
1	−1	1	−3	1
1	1	−1	9	False
1	1	1	−3	1

NAND

$f = \overline{x_1 x_2}$

$E = -v_1 - v_2 - 2v_3 + v_1 v_2 + 2v_1 v_3 + 2v_2 v_3$

v_1	v_2	v_3	E	f
−1	−1	−1	9	False
−1	−1	1	−3	1
−1	1	−1	1	False
−1	1	1	−3	1
1	−1	−1	1	False
1	−1	1	−3	1
1	1	−1	−3	−1
1	1	1	1	False

NOR

$f = \overline{x_1 \vee x_2}$

$E = v_1 + v_2 + 2v_3 + v_1 v_2 + 2v_1 v_3 + 2v_2 v_3$

v_1	v_2	v_3	E	f
−1	−1	−1	1	False
−1	−1	1	−3	1
−1	1	−1	−3	−1
−1	1	1	1	False
1	−1	−1	−3	−1
1	−1	1	1	False
1	1	−1	−3	−1
1	1	1	9	False

EXOR

$f = x_1 \oplus x_2$

$E = v_1 + v_2 + v_3 - 2i + v_1 v_2 + v_1 v_3 + v_2 v_3 - 2i v_1 - 2i v_2 - 2i v_3$

v_1	v_2	v_3	I	E	f
−1	−1	−1	−1	−4	−1
−1	−1	−1	1	4	False
−1	−1	1	−1	−2	False
−1	−1	1	1	−2	False
−1	1	−1	−1	−2	False
−1	1	−1	1	−2	False
−1	1	1	−1	4	False
−1	1	1	1	−4	1
1	−1	−1	−1	−2	False
1	−1	−1	1	−2	False
1	−1	1	−1	4	False
1	−1	1	1	−4	1
1	1	−1	−1	4	False
1	1	−1	1	−4	−1
1	1	1	−1	14	False
1	1	1	1	−2	False

5.2   EXPERIMENTS

In this section, we introduce experiments with discrete Hopfield models under various levels of noise, that was injected in the form of uncorrected states of threshold elements. We are interested in the number of iterations of such a noisy model required to converge on a correct result.

5.2.1   METRIC

Discrete noise is added with some probability to each threshold gate of a Hopfield model. Noise is measured by *noise probability*, which is defined as the probability that a neuron cell is corrupted by noise resulting in a bit flip, called "a state flip" in terms of Hopfield networks. In other words, a change of state is modeled by a uniformly distributed discrete random signal. For example, a noise level of 0.1 (10%) means that the probability of changing the current state of the neuron cell is 0.1.

5.2.2   UPDATING

There are two updating rules that apply to the Hopfield network: (a) the deterministic rule, and (b) the stochastic Boltzmann rule [42]. The Boltzmann updating rule is based on the assumption of uncertainty about the state of a particular cell during the updating process. Instead of setting the state of cell k deterministically, the process uses the probability $p(k)$ that cell k takes state 1. This involves the following steps: (a) Select the control parameter T; (b) Randomly select a cell k, (c) Calculate $\Delta E(k) = \sum_i^n x_i c_{ik} - N_k$. If $\Delta E(k) > 0$, then calculate the probability that cell k takes state 1: $p(k) = \frac{1}{1+\exp(-\Delta E(k)/T)}$. Otherwise, set the state of cell k to -1; (d) Repeat steps (b) through (d) until the state of cell does not change for a certain interval (this is also called the stable state condition); for example, 20 iterations.

5.2.3   NETWORKING HOPFIELD MODELS OF GATES

In our experiments, we used Hopfield models of two-input and three-input logic gates, which corresponds to networks of 3-, 4-, and 5-threshold gates such as those shown in Fig. 5.4.

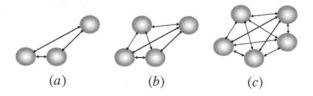

(a) (b) (c)

Figure 5.4: Configurations of threshold elements in Hopfield models: (a) three-node model (for a NOT gate and the two-input logic gates, except an EXOR gate), (b) four-node model (for the three-input logic gates and a two-input EXOR gate), and (c) five-node model.

In this section, we introduce a technique for mapping logic networks (connected logic gates) into networks of Hopfield models (connected Hopfield models of logic gates). We use a technique that was initially introduced in [18]. There are two merging rules for the connection of Hopfield models.

Rule I: Hopfield gate merging

Connecting an output of one Hopfield model of a logic gate to the input of another (cascading the gates) is performed by merging the corresponding "output" cell of the first gate with the "input" cell of the second one. The threshold value of the new cell is the cumulative value of thresholds of the merged cells.

Rule II: Hopfield gate merging

Connecting an output of one Hopfield model of logic gate to the input of another (cascading the gates), while the second input of the other gate is also the input of the first one, is performed by merging the corresponding "output" cell of the first gate with the "input" cell of the second one, as well as both the "input" cells. The threshold value of the two new cells is the cumulative threshold of the merged cells.

For example, accordingly to the Rule I, while connecting an AND and EXOR gate, both cells $G1$ are merged in a new cell, with threshold value 2+1=3, obtained by adding the threshold values of the merged cells (Fig. 5.5, upper plane). Application of Rule II is shown in the lower plane of Fig. 5.5, where input x_2 is common to both AND and EXOR gates, so their cascading involves the merging of nodes $X2$ and $G1$, and their thresholds are the cumulative values -1+1=0 and 2+1=3, respectively.

The design taxonomy presented in Fig. 5.5 is only useful for small networks containing a few gates. This is because the relaxation time increases drastically as the networks are scaled up. Hopfield gates (models) can be connected into a network using I/O interfaces implemented by so-called *hidden cells* [27, 40].

5.2.4   MODELING AND RESULTS

We compared the robustness of Hopfield models for EXOR gates in terms of the number of iterations, and the probability of achieving a stable state with respect to the level of noise injected into threshold elements. In the experiment, we use three possible implementations of the two-input EXOR gate: (a) a direct modeling of a two-input EXOR gate by a four-node Hopfield model with Boltzmann updating rule (Table 5.2); (b) a network of Hopfield models for AND, NOT, and OR gates (since EXOR $= \overline{x}y \vee x\overline{y}$), and (c) a network of Hopfield models for NAND, AND, and OR gates (since EXOR $= \overline{(x \wedge x)y \vee x\overline{(y \wedge y)}}$).

The results of these experiments are given in Fig. 5.6. We observe that EXOR networks of greater complexity result in better noise-tolerance properties. For example, given a noise probability at a node ranging between 0 and 0.5, the number of iterations required to reach a stable state varies between the EXOR Hopfield gate itself (A-implementation) (Fig. 5.6a), and the implementation of an EXOR gate using a network of NOT, AND, and OR gates (B-implementation), as well as a network of NAND, AND, and OR gates (C-implementation). The latter two seemed to need fewer iterations after the noise level increased beyond 0.2, and substantially fewer iterations beyond

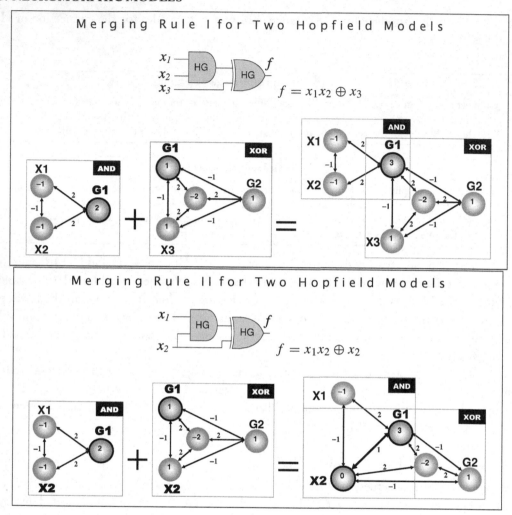

Figure 5.5: Connection of two Hopfield Gates (the Hopfield models of logic gates) denoted by HG.

a noise level of 0.5. The EXOR gate itself (which requires four threshold cells) required many more iterations to achieve the correct output at high noise level, while the networks of NOT, AND, OR, or NAND gates, each consisting of two (NOT only) or three (threshold cells), connected in the network with total 10 threshold cells (B−type) and 15 threshold cells (C−type), respectively, to reach the stable state mush faster.

Given a node noise probability ranging between 0 and 0.5, the probability of achieving the correct output is better for the EXOR Hopfield gate itself (A−type) (Fig. 5.6b), and for the implementation of an EXOR gate on a network of NAND, AND, and OR gates (C−type), while the

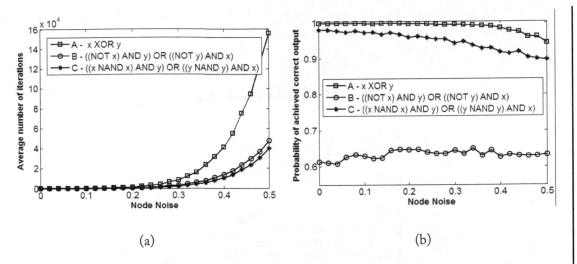

(a)   (b)

Figure 5.6: (a) Comparison of Hopfield models for various implementations of a two-input EXOR gate, in terms of the number of iterations with respect to the noise level: (A) four-node Hopfield model of EXOR gate, (B) Hopfield model of EXOR as $f = \bar{x}y \vee x\bar{y}$, and (C) the Hopfield model of EXOR gate given by $f = \overline{(x \wedge x)}y \vee x\overline{(y \wedge y)}$; (b) Comparison of Hopfield models for various implementations of a two-input EXOR gate in terms of their probability of achieving a correct result with respect to the applied noise level.

probability of achieving the correct output for the network of NOT, AND, and OR gates ($B-$type) is only around 0.6.

Another experimental study of the performance of the Hopfield networks, implementing simple AND-EXOR expressions, was performed using the circuits $x_1x_2 \oplus x_3x_4$. This experiment compared the probability of achieving the correct output of the benchmark, for both a logic network and a Hopfield network with noise. Fig. 5.7 shows the probability of correct output given a noise probability ranging from 0 to 0.5. The stable state condition for the Hopfield network is set to 15, 10, and 5 iterations. The results show that the probability of achieving the correct output is higher for the Hopfield network compared to the logic network, as the noise probability increases. For example, given the noise probability of 0.2, the Hopfield network converges to a stable state condition in 10 iterations, thus achieving the correct output with probability 0.96, while the logic network is only able to achieve the correct output with probability 0.81 (a difference of 15%). It is also observed that by increasing the number of iterations (for the stable state condition) from 10 and to 15, the Hopfield network is able to achieved a probability of correct output $> 99\%$. This shows that Hopfield implementation is capable of operating with high accuracy in noisy environments.

These results confirm that the behavior of Hopfield networks and their modifications, such as Boltzmann machines, demonstrates high fault tolerance in the presence of critical noise in a part or in some or all neurons of the network.

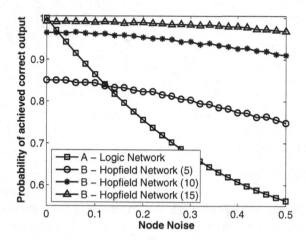

Figure 5.7: Behavioral characteristics of a conventional logic network, $f = x_1 x_2 \oplus x_3 x_4$ given a noise level, compared with its Hopfield model (B) (with Boltzmann updating) given a fixed number of iterations (5, 10, and 15).

5.3 MULTISTATE HOPFIELD MODEL

A k-state Hopfield model of a k-valued logic gate consists of fully connected k-threshold gates, with each gate operating in complex domain [51, 132, 160]. In this model, the values of a k-valued logic gate are encoded in the complex domain by the value exp $\frac{i2\pi j}{k}$, for $j \in \{0, 1, 2, \ldots k - 1\}$, $i = \sqrt{-1}$. For example, for a 2-state Hopfield model ($k = 2$ and $j = 0, 1$), the values of a switching function are encoded as exp $\frac{i2\pi j}{2} = $ exp $0 = 1$, if $j = 0$, and exp $i\pi = -1$, if $j = 1$. The values of the ternary function ($k = 3$ and $j = 0, 1, 2$) are encoded as 1, -1, and 0.

5.4 CONCLUDING REMARKS

The Hopfield network and its extension, the Boltzmann machine, used for modeling noise-tolerant logic gates and networks, exploit an approach similar to the MRF computational and modeling techniques, such as embedding logic functions via a compatibility table, Gibbs joint probability, and maximum likelihood estimation, that are similar to the MRF approach. However, locality of computing is an inherent property of the Hopfield model because it is a complete graph model by definition, but neighboring must be specified in the MRF model. In this chapter, we have introduced a preliminary experimental study of discrete Hopfield and Boltzmann models for logic gates. We found that these models, like the MRF models, are good candidates for hardware implementation of noise-tolerant logic gates. Specifically, we came to the following conclusions:

1. Preliminary comparison of the Boltzmann model and the MRF model shows that they result in similar noise-tolerance properties for logic gates. However, Boltzmann and MRF models differ in hardware implementation.

2. The Boltzmann model is sensitive even to the details of gate implementation. This conclusion follows from our study of noise-tolerant EXOR gate implementation (Fig. 5.6). However, these differences can be overcome by using the same computing data structure as decision diagrams.

3. The Hopfield model can be generalized for multivalued logic functions. To do this, binary threshold cells must be replaced by their multi-threshold extensions. A number of benefits are expected from this model. Theoretically, it is possible to decrease power dissipation and interconnections as compared with s binary mode using this method.

4. An arbitrary Hopfield and Boltzmann network can be represented by a decision diagram. Decision diagrams have various valuable features for implementation; in particular, they are switch-based structures and that can be embedded into a 3D topology. This is a perspective study related to a particular technology. The Boltzmann model and the MRF model can be represented, compared, and implemented in this form of data structure.

CHAPTER 6

Noise-tolerance via error correcting

Murphy's law

Regulation is the substitution of
error for chance

6.1 INTRODUCTION

The use of coding methods for error correction is an integral part of nano devices design. This chapter introduces the design philosophy of correcting rather than avoiding errors. In this approach, error-protection bits are added to the information-bearing data bits using the data structures such as Binary Decision Diagrams (BDDs). In this chapter, we are based on the BDD-based error correction techniques introduced by Astola et. al [8]. The term "error" is referred to as an undesirable event affecting the information content of logical signals, due to a failure in the physical level and/or a fault in the logical level of the system.

6.2 BACKGROUND OF ERROR CORRECTING TECHNIQUES

Error-correcting codes were originally intended to efficiently protect the integrity of data communication systems. This protection technique implies maximizing the probability of detecting/correcting errors in the transmitted data while minimizing the required hardware overhead. In this approach, a computing system has been considered as a data transmission channel [85, 107, 108].

Terminology. Coding is the representation of data by *code symbols*, or *sequences*, of code symbols. Data is placed into the code form by *encoding*, and is extracted from the code form by *decoding*. The number of code symbols in the encoded data is called a *code length*. Codes can have a large average code length. In error-control coding, the code length always exceeds the one required for the unique representation of data. Such codes are said to contain *redundancy*. *Error detection* and *error correction* techniques require additional, or *redundant bits*, to be appended to the data, or information bits.

The repetition encoding.   A simple way to correct as well as detect errors is to repeat each bit a certain number of times. The assumption is that the expected bit occurs more often. The scheme can tolerate error rates up to one error in every two bits transmitted at the expense of the increased bandwidth. Consider a stream of 0s and 1s to be sent. This data is grouped into blocks of bits. Let the stream be 1011. By repeating this block three times each, the blocks 1011 1011 1011 are sent. Suppose that the blocks 1010 1011 1011 are received. One group is not the same as the other two, and it is concluded that an error has occurred. This scheme is not efficient, because it can only detect double errors, or correct single errors that occurred in exactly the same place for each group. For example, errors cannot be detected if 1010 1010 1010 is received.

Von Neumann's NAND gate repetition encoding.   The repetition scheme was studied by von Neumann in his work on the correction of the operation of NAND function [143]. The scheme involves replicating the function to be multiplexed N times. N wires are used to carry the signal of each input and output. Von Neumann demonstrated that this multiplexing technique based on repetition encoding works in cases where elements have uniform probability of failure less than $p = 0.0107$. For example, to achieve an overall failure probability of 0.027 for a single multiplexed NAND function, $N = 1000$ for $p = 0.005$. Von Neumann proved that the complexity of correction and redundancy using repetition encoding at the gate level are so high that it makes redundancy techniques useless. However, recent revival of this simplest encoding is motivated by the fact that self-assembling nanostructures can potentially reach the redundancy levels, required for reliable computing. Note that redundancy is essential for reliable computation because helps combat noise.

Single error-detecting codes.   A single error-detection code is capable of detecting, but not correcting the single errors. The simplest is the *parity* code. This code is formed by adding one check symbol to each block of k information symbols. The added symbol is called a *parity check symbol* and is used as follows: (*a*) If the received word contains an *even* number of errors, the decoder will not detect the errors. (*b*) If the number of errors is *odd*, the decoder will detect that an odd number of errors, most likely one, has occurred.

Error-correcting coding.   The origin of error-control coding goes back to 1948 when Claude Shannon presented the fundamental theorem of information theory. The theorem states that if the rate of data transmission through a noisy channel is less than the capacity of the channel, there exist codes that can make the probability of error at the receiver arbitrarily small. The theorem does not provide for techniques to synthesize these codes; it only shows how to analyze the quality of the code. The best near Shannon limit error-correcting coding is proposed in [11]. Application of this technique for the Hopfield model has been studied in [12].

An example of error-correcting code is a Hamming code. The Hamming code is characterized by a *distance n*, which is the number of the bits needed to be flipped to obtain another code word with no visible errors. A (3,1) repetition code (each bit is repeated three times) has a distance of 3, as three bits need to be flipped in the same triple to obtain another code word with no visible errors.

A (4,1) repetition code has a distance of 4, so flipping two bits can be detected, but not corrected; flipping three bits in the same group can lead to correcting toward the wrong code word.

In the single error-correcting Hamming code, information bits along with several parity bits compose a codeword. The values of the parity bits are determined by an even-parity scheme over selected information bits. After a codeword has been transmitted, the parity bits are recalculated at the receiver end to check whether the correct parity still exists over the selected information bits. By comparing the recalculated parity bits against those in the received word, it is possible to determine if the received codeword is free of single errors, if a single error has occurred, and exactly which bit has erroneously changed.

If more than one bit is changed during transmission, then this coding scheme is no longer capable of determining the location of the errors. A single error in a codeword will cause one or more parity checks to fail. The parity check pattern can be used to locate the position of error, provided the checking domains of each parity bit have been chosen correctly. Determining the error position in a binary message is sufficient for correction; the erroneous digit must be readily flipped, or inverted.

Hamming codes work well in situations where one can reasonably expect errors to be rare events. For example, some memory devices have error ratings in the order of 1 bit per 100 million. A distance 3 Hamming code is capable of correcting this type of errors. However, it is useless in situations when there is a likelihood that multiple random bit errors may occur. In situations such as this, codes with more complicated structures are exercised for error control.

Example of the 7-bit Hamming code.   Given four information bits, three parity bits are included along with the four information bits to form a seven-bit codeword. The structure of the codeword in this case is given in Figure 6.1. The following rule is used for the parity bits p_1, p_2, and p_3:

$p_1 = 0$ if $b_1 \oplus b_2 \oplus b_4$, and $p_1 = 1$ otherwise.

$p_2 = 0$ if $b_1 \oplus b_3 \oplus b_4$, and $p_2 = 1$ otherwise.

$p_3 = 0$ if $b_2 \oplus b_3 \oplus b_4$, and $p_3 = 1$ otherwise.

Systematic generator matrix.   Hamming codes are linear codes which can be defined by a generator matrix and parity-check matrix in systematic form. In a *systematic* code, k digits of the codeword coincide with the message digits, and the remaining $r = n - k$ digits are parity digits. Such a structure simplifies the decoding procedure. A generator matrix is called *systematic* if it has the form as follows:

$$G = \begin{bmatrix} P & | & I_k \end{bmatrix} = \left[\begin{array}{cccc|cccc} p_{11} & p_{12} & \cdots & p_{1r} & 1 & 0 & \cdots & 0 \\ p_{21} & p_{22} & \cdots & p_{2r} & 0 & 1 & \cdots & 0 \\ \vdots & & & & & & \vdots & \\ p_{k1} & p_{k2} & \cdots & p_{kr} & 0 & 0 & \cdots & 1 \end{array} \right] \tag{6.1}$$

where P is the parity check portion of G, and I_k is the $k \times k$ identity matrix.

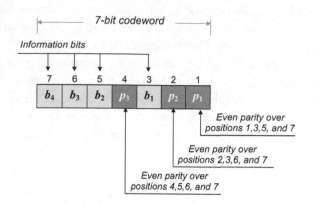

Figure 6.1: The structure of the seven-bit Hamming code.

Parity-check matrix.   For a $k \times n$ generator matrix G, there exists an $r \times n$ matrix H such that $GH^T = 0$, where H^T is the *transpose* of H. This matrix, called a *parity-check matrix*, is represented in the following form:

$$H = \begin{bmatrix} I_r \mid P^T \end{bmatrix} = \begin{bmatrix} 1 & 0 & \cdots & 0 & p_{11} & p_{21} & \cdots & p_{k1} \\ 0 & 1 & \cdots & 0 & p_{12} & p_{22} & \cdots & p_{k2} \\ \vdots & & & & & & & \\ 0 & 0 & \cdots & 1 & p_{1r} & p_{2r} & \cdots & p_{kr} \end{bmatrix} \tag{6.2}$$

6.3   LOGIC GATE RELIABILITY

Deterministic description of a logic gate is reasonable when the gate operates on noise-free signals, that is, when noise can be neglected. Example of a noise-free AND gate which operates with binary signals, $x_1, x_2 \in \{0, 1\}$, and produces the binary signal, $f = x_1 \wedge x_2$, $f \in \{0, 1\}$, according to its truth table.

Reliability analysis of logic gates refers to the problem of evaluating the effects of errors due to various sources of noise. Typical techniques for reliability analysis utilize noise (fault) injection and simulation in a Monte Carlo framework. The gate noise and line (or wire) noise are usually considered.

Gate error.   The *output probability* of a logic gate is defined as a probability that the function will assume the value 1 given the probabilities that each of the input variables are assigned the value 1. For example, assuming all inputs have equal probability of being 0 or 1, (a) the output probability of a 2-input AND gate is 1/4, (b) the output probability of a 3-input OR gate is 7/8, and (c) two-input NAND gate has output probability 3/4.

Von Neumann viewed a noisy logic gate as computing a Boolean function (correctly), and then complementing the result with probability p. Noise was considered as a single random event which is independent of the input to the gate. The system was considered reliable if the probability of its correct output is greater than a certain threshold. In von Neumann's multiplexing scheme, the noisy NAND gates were assumed to flip the output signal values with the probability of $p \leq 1/2$, while the input and output lines function reliably. If gates of a logic circuit are too noisy, then the circuit cannot be used to compute functions reliably [143].

Following von Neumann's view, we define the logic gate reliability as the probability, R, that the gate will correctly perform its logic operation: $R = 1 - p_g$, where p_g is the gate error probability. Therefore, the gate error probability is defined as the probability that the gate output is inverted, or flipped.

Discrete channel. The feasibility of reliable data transmission over noisy communication links was derived by C. Shannon in 1948. Being applied to data transmission between logic gates in a circuit, Shannon information theory states that it is possible to compute reliably using a logic gate which is subject to a random or unpredictable noise [146]. Also, it is possible to determine the minimum energy needed to compute a task in the presence of noise.

In the classical von Neumann model for errors, noise at a gate is modeled as a binary symmetric channel, with a crossover probability p [143]. Also, von Neumann employed the simplest possible coding scheme called n-repetition code, where the same information bit is transmitted n times, and majority logic is used at the receiving end for decoding the message bit.

A graphical model of a discrete probabilistic channel in a communication system is shown in Fig. 6.2(a). The channel is defined as a conditional probability distribution $p(\mathbf{y}$ received $\mid \mathbf{x}$ transmitted) over a finite input alphabet F, and a finite output alphabet Φ, for every pair $(\mathbf{x}, \mathbf{y}) \in F^m \times \Phi^m$, where F^m and Φ^m denote the set of all words of length m over F and Φ, respectively. For example, given ($F = \Phi = \{0, 1\}, \mathbf{x} = x_1, x_2, \ldots, x_m$ and $\mathbf{y} = y_1, y_2, \ldots, y_m$,

$$p(\mathbf{y} \text{ received} \mid \mathbf{x} \text{ transmitted}) = \prod_{j=1}^{m} p(y_i \text{ received} \mid x_i \text{ transmitted}),$$

where $p(y$ received $\mid x$ transmitted$) = \begin{cases} 1 - p, & \text{if } y = x; \\ p, & \text{if } y \neq x. \end{cases}$ and p is called a crossover probability of the channel. The probability p can be calculated from a knowledge of the signals used and the probability distribution of noisy signals. The action of the channel can be described as flipping each input bit with probability p independently. The channel is called symmetric since the probability of the flip is the same regardless of whether the input is 0 or 1. The graph in Fig. 6.2(a) indicates that the probability of the output being correct is $(1 - p)$, and the probability that it is incorrect is p, for all inputs. This graph can be interpreted as a model of a noisy noninverting buffer.

Discrete channel model of logic gates. A noisy NOT gate and a noisy 2-input logic gate of an arbitrary type (OR, AND, NOR, NAND, EXOR) is represented in Fig. 6.2(b) and Fig. 6.2(c), respectively. The cases $p = 0$ and $p = 1$ correspond to reliable communication, given an error-free gate, the probability crossover $p = 1/2$ stands for the case when the output of the channel is statistically independent of its input, that is, corresponds to noisy gate. The channel model can be used to represent the output of a gate, or the node of a decision diagram. The channel switches the output symmetrically from 0 to 1 or from 1 to 0 with the same probability of error, $p \in \{0, 0.5\}$. The scenario when $p > 0.5$ means that the output of the gate is more likely to be faulty than correct. This is the motivation to consider the gate failure probability in the range between 0 and 0.5.

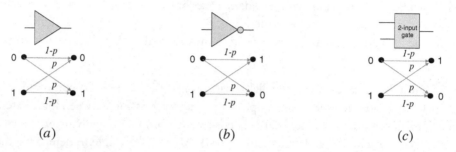

(a)   (b)   (c)

Figure 6.2: Discrete probabilistic channel model for (a) noninverting buffer, (b) the NOT gate, and 2-input logic gate.

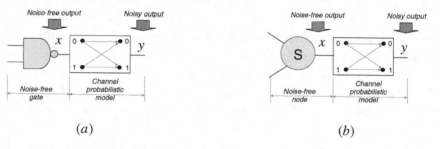

(a)   (b)

Figure 6.3: A channel model of a noisy gate (a), and node (b).

Probability expressions. The following technique relates Boolean operations on Boolean (deterministic) variables to the corresponding operations on random variables (probabilities). Let $f = f(x_1, x_2, \ldots, x_n)$ be a Boolean function. Denote the probability of an input signal x_i being 1 as $p(x_i = 1) = X_i, i = 1, 2, \ldots, n$. The *probability* expression of f is defined as $p(f = 1) = F(X_1, X_2, \ldots, X_n) = F$. The probability expression of a Boolean function is related to an *arithmetic* expression or *pseudo Boolean* representation [125, 153, 154]. The following rules are applied in manipulation of probability expressions [103, 144]:

<u>Rule 1</u>: $p(\overline{x}_i = 1) = 1 - X_i$,

<u>Rule 2</u>: $p(x_i \wedge x_i = 1) = X_i$,

<u>Rule 3</u>: $p(x_i \wedge x_j = 1) = X_i X_j$, $i \neq j$ (for independent X_i and X_j), and

<u>Rule 4</u>: $p(f_1 \vee f_2 = 1) = F_1 + F_2 - F_1 F_2$, where $F_i = p(f_i = 1)$.

For example, the probability expression for a 2-input OR gate, $f = x_1 \vee x_2$, is derived using the above rules as follows: $p(x_1 \vee x_2 = 1) = p(\overline{x_1 \overline{x_2}} = 1) = 1 - (1 - X_1)(1 - X_2)$. Let the inputs x_1 and x_2 be random variables with probabilities $p(x_1)$ and $p(x_2)$, respectively. Then the output probability, $p(f = 1)$, of OR gate is $p(f = 1) = 1 - (1 - p(x_1))(1 - p(x_2))$. Let $p(x_1) = p(x_2) = p$, then $p(f = 1) = p(2 - p)$. By analogy, the probability expression for a Boolean function $f = \overline{x}_1 x_2 \vee x_2 x_3$ is $F = p(f = 1) = X_2 - X_1 X_2 + X_1 X_2 X_3 = p(x_2) - p(x_1)p(x_2) + p(x_1)p(x_2)p(x_3)$.

In Table 6.1, the relation of Boolean operations (with Boolean variables) and probability operations (with random variables) is introduced. Each binary signal of a gate input or output is associated with a random variable, which denotes the probability of this signal. The probabilistic expressions are derived using both algebraic and tabular forms of logic operations. It is assumed that the probability of any input signal being 1 is $p(x_1) = p(x_2) = p$, and the input signals are independent. It is implied that $p(f = 1) + p(f = 0) = 1$.

6.3.1 MULTIPLEXER CIRCUIT RELIABILITY.

A circuit on 2-to-1 switches, or simple multiplexers, is an alternative way (compared to the a network of logic gates) to implement switching functions. It is formally described by Binary Decision Trees (BDTs) or by its reduced form, Binary Decision Diagrams (BDDs).

Binary decision diagrams. A node of a BDD corresponds to a Shannon expansion with respect to a variable x_i, and can be implemented as a bi-directional 2-to-1 multiplexer, which control input represents this variable.

The Shannon expansion of a function f with respect to a variable x_i is derived as follows:

$$ f = \overline{x}_i f_0 \vee x_i f_1 = \overline{x}_i f_0 + x_i f_1 - (\overline{x}_i f_0)(x_i f_1) = \overline{x}_i f_0 + x_i f_1 $$

where $f_0 = f_{x_i=0}$, $f_1 = f_{x_i=1}$, and $(\overline{x}_i f_0)(x_i f_1) = 0$. Assuming an independency of the co-factors $\overline{x}_i$ and f_0 and x_i and f_1, the Shannon expansion in terms of function probabilities is written as follows:

$$ \begin{aligned} p(f) &= p(\overline{x}_i f_0) + p(x_i f_1) = p(\overline{x}_i)p(f_0) + p(x_i)p(f_1) \\ &= p(x_i = 0)p(f_{x_i=0}) + p(x_i = 1)p(f_{x_i=1}) \end{aligned} \tag{6.3} $$

In [154] (chapter 20), two approaches to evaluate output probabilities are introduced: (a) a top-down approach by Thornton and Nair [133] and (b) bottom-up approach by Miller.[1] We follow the bottom-up algorithm as shown in Fig. 6.4. This approach can be extended to shared BDDs, computation spectral coefficients, and cross-correlation of functions [154].

A node in a BDD, which is assigned with the i-th variable, x_i, and with outgoing edges corresponding to $f_{x_i=0}$ and $f_{x_i=1}$, computes the Shannon expansion $f = \overline{x}_i f_0 \vee x_i f_1$, where $f_0 =$

---

[1]To the best of our knowledge, Shannon expansion in terms of probabilities for BDDs have been introduced by Krieger [65]

Table 6.1: Relation between the operations on Boolean and random variables.

Gate	Boolean operation	Probability operation

NOT gate ($x \rightarrow f$)

(a) algebraic form: $f = \bar{x}$
(b) tabular form

x	f
0	1
1	0

(a) algebraic form:
$p(f = 1) = 1 - p(x),$
$p(f = 0) = 1 - p(f = 1) = p(x)$
(b) tabular form

$p(x)$	$p(f)$
$p(0)$	$1 - p = p(f = 1)$
$p(1)$	$p = p(f = 0)$

AND gate ($x_1, x_2 \rightarrow f$)

(a) algebraic form: $f = x_1 \wedge x_2$
(b) tabular form

x_1	x_2	f
0	0	0
0	1	0
1	0	0
1	1	1

For independent input signals:
(a) algebraic form:
$p(f = 1) = p(x_1)p(x_2) = p^2,$
$p(f = 0) = 1 - p(f = 1) = 1 - p^2$
(b) tabular form

$p(x_1)$	$p(x_2)$	$p(f)$
$p(0)$	$p(0)$	$(1 - p)^2$
$p(0)$	$p(1)$	$+(1 - p)p$
$p(1)$	$p(0)$	$+p(1 - p) = 1 - p^2 = p(f = 0)$
$p(1)$	$p(1)$	$p^2 = p(f = 1)$

OR gate ($x_1, x_2 \rightarrow f$)

(a) algebraic form: $f = x_1 \vee x_2$
(b) tabular form

x_1	x_2	f
0	0	0
0	1	1
1	0	1
1	1	1

(a) algebraic form:
$p(f = 1) = p(2 - p),$
$p(f = 0) = 1 - p(f = 1) = (1 - p)^2$
(b) tabular form

$p(x_1)$	$p(x_2)$	$p(f)$
$p(0)$	$p(0)$	$(1 - p)^2 = p(f = 0)$
$p(0)$	$p(1)$	$(1 - p)p$
$p(1)$	$p(0)$	$+p(1 - p)$
$p(1)$	$p(1)$	$+p^2 = p(2 - p) = p(f = 1)$

NAND gate ($x_1, x_2 \rightarrow f$)

(a) algebraic form: $f = \overline{x_1 x_2}$
(b) tabular form

x_1	x_2	f
0	0	1
0	1	1
1	0	1
1	1	0

(a) algebraic form:
$p(f = 1) = 1 - p^2,$
$p(f = 0) = p^2$
(b) tabular form

$p(x_1)$	$p(x_2)$	$p(f)$
$p(0)$	$p(0)$	$(1 - p)^2$
$p(0)$	$p(1)$	$+(1 - p)p$
$p(1)$	$p(0)$	$+p(1 - p) = 1 - p^2 = p(1)$
$p(1)$	$p(1)$	$p^2 = p(f = 1)$

$f_{x_i=0}$ and $f_1 = f_{x_i=1}$. Assuming that random variables $\bar{x}_i$ and f_0, and x_i and f_1 are independent, probability is propagated from inputs to the output as follows: $p(f) = p_{x_i=0} p_{f_{x_i=0}} + p_{x_i=1} p_{f_{x_i=1}}$ [123, 133, 134].

Information-theoretical measures. The reliability estimation of computing in switching networks was considered in [2, 7, 122, 141] using information-theoretical approach. Consider, for example, a 2-input AND function with four random combinations of input signals: 00 with probability p_{00}, 01 with probability p_{01}, 10 with probability p_{10}, and 11 with probability p_{11} (Figure 6.5a). The

Probability propagation in decision tree

Node operation: The Shannon expansion in terms of probabilities,
$$p(f) = p(x_i = 0)p(f_{x_i=0}) + p(x_i = 1)p(f_{x_i=1})$$
Initial condition: Assign a probability of 1 to the terminal node 1, $p(1) = 1$, and a probability of 0 to the terminal node 0, $p(0) = 0$.

Initial sets: $p(f_{x_i=1}) = p$ and $p(f_{x_i=0}) = 1 - p$

Step 1: Calculate the probabilities $p(\overline{x}_1)$ and $p(x_1)$:

$$
\begin{aligned}
p(\overline{x}_1) &= p(x_2 = 0)p(f_{x_2=0}) + p(x_2 = 1)p(f_{x_2=1}) \\
&= 1 \times (1 - p) + 1 \times p = 1 \\
p(x_1) &= p(x_2 = 0)p(f_{x_2=0}) + p(x_2 = 1)p(f_{x_2=1}) \\
&= 1 \times (1 - p) + 0 \times p = 1 - p
\end{aligned}
$$

Step 2: Calculate the output probability $p(f)$:

$$
\begin{aligned}
p(f) &= p(x_1 = 0)p(f_{x_1=0}) + p(x_1 = 1)p(f_{x_1=1}) \\
&= 1 \times (1 - p) + (1 - p) \times p = \boxed{1 - p^2}
\end{aligned}
$$

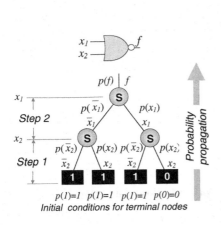

Probability propagation in BDD

Step 1: Calculate the probability $p(x_1)$:

$$
\begin{aligned}
p(\overline{x}_1) &= p(x_2 = 0)p(f_{x_2=0}) + p(x_2 = 1)p(f_{x_2=1}) \\
&= 1 \times (1 - p) + 0 \times p = 1 - p
\end{aligned}
$$

Step 2: Calculate the output probability $p(f)$:

$$
\begin{aligned}
p(f) &= p(x_1 = 0)p(f_{x_1=0}) + p(x_1 = 1)p(f_{x_1=1}) \\
&= 1 \times (1 - p) + (1 - p) \times p = \boxed{1 - p^2}
\end{aligned}
$$

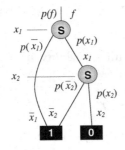

Figure 6.4: (a) Probability propagation in a decision tree for representation of a noisy 2-input NAND gate; (b) probability propagation in the BDD representation of a noisy 2-input NAND gate.

entropy of the input signals is calculated as follows:

$$
\begin{aligned}
H_{in} = &- p_{00} \times \log_2 p_{00} - p_{01} \times \log_2 p_{01} \\
&- p_{10} \times \log_2 p_{10} - p_3 \times \log_2 p_{11} \; \textit{bit/pattern}
\end{aligned}
$$

The maximum entropy of the input signals can be calculated by inserting into the above equation $p_i = 0.25$, $i = 0, 1, 2, 3$. For example, the output of the AND function is equal to 0 with probability 0.75, and to 1 with probability 0.25. The entropy of the output signal is $H_{out} = -0.25 \times \log_2 0.25 - 0.75 \times \log_2 0.75 = 0.81 \; \textit{bit/pattern}$.

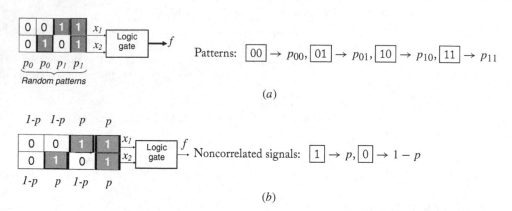

Figure 6.5: Measurement of probabilities: random patterns (a) and uncorrelated signals (b).

Let the input signal be equal to 1 with probability p, and 0 with probability $(1 - p)$ (Fig. 6.5b). The entropy of the input signal is

$$
\begin{aligned}
H_{in} &= -(1 - p)^2 \times \log_2(1 - p)^2 - 2(1 - p)p \times \log_2(1 - p)p - p^2 \times \log_2 p^2 \\
&= -2(1 - p) \times \log_2(1 - p) - 2p \times \log_2 p \quad bit
\end{aligned}
$$

For example, the output of the AND function is equal to 1 with probability p^2, and to 0 with probability $1 - p^2$. Hence, the entropy of the output signal is

$$
H_{out} = -p^2 \times \log_2 p^2 - (1 - p^2) \times \log_2(1 - p^2) \quad bit
$$

The maximum value of the output entropy is equal to 1 when $p = 0.707$. Hence, the input entropy of the AND function is 0.745 *bits*. Thus, in the case of uncorrelated signals, information losses are less.

Fig. 6.6 illustrates the calculation of entropy on a BDT. In the top-down BDT design, two information measures are used: conditional entropy, $H(f|\text{Tree})$, and mutual information, $I(f; \text{Tree})$. The initial state of this process is characterized by the maximum value for the conditional entropy $H(f|\text{Tree}) = H(f)$. Nodes are recursively attached to the BDT by using the top-down strategy. In this strategy the entropy $H(f|\text{Tree})$ of the function is reduced, and the information $I(f; \text{Tree})$ increases, since the variables convey information about the function. Each intermediate state can be described in terms of entropy by the equation $I(f; Tree) = H(f) - H(f|\text{Tree})$. The goal of such an optimization is to maximize the information $I(f; \text{Tree})$ that corresponds to the minimization of entropy $H(f|\text{Tree})$, in each step of the BDT design. The final state of the BDT is characterized by $H(f|\text{Tree}) = 0$ and $I(f; \text{Tree}) = H(f)$, that is, Tree represents the Boolean function f.

The BDT and BDD design process is a recursive Shannon decomposition of a Boolean function f. A step of this decomposition corresponds to the expansion of f with respect to a variable x. Assume that the variable x in f conveys information that is, in some sense, the rate of

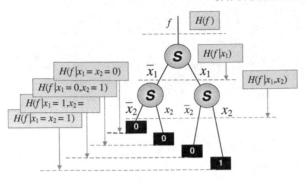

Figure 6.6: Measurement of entropy on a BDT.

influence of this variable on the output value for f. The information measure of Shannon expansion, S, for a Boolean function f with respect to a variable x is represented by the equation

$$H(f|x) = p_{x=0}H(f_{x=0}) + p_{x=1}H(f_{x=1})$$

Details of application of information-theoretical estimation to decision diagrams can be found in [19, 20, 154].

6.4 NOISE MODELING IN BDDS.

In [58, 60], BDDs are mapped to nanoscale structures, such as the hexagonal BDD quantum node devices. Correct operation of such devices at nanoscale requires mitigation of two distinct sources of faults. The first source is noise, as the signal levels are extremely low. The second source is incorrect switching at the nodes or missing wiring due to defects [67].

2-input gates. Consider a BDD representation of a NAND gate, inputs of which are represented by random variables. The goal of the noise simulation is to analyze the circuit response and to compare the theoretical estimation (see Fig. 6.4) against the Monte Carlo simulation. In Monte Carlo simulation, the uniformly distributed random variables were generated, given the probability of error at the inputs x_1 and x_2, and the result was averaged over a thousand runs. These two approaches are shown in Fig. 6.7. The simulation showed that the BDD of the NAND gate has better reliability than a regular NAND gate. Monte Carlo simulation closely matched the theoretical estimation. The similar results were obtained for the AND and OR gates.

Input error probability and SNR. The signal-to-noise-ratio (SNR) is a measure of system performance, defined as the ratio of a signal power to an unwanted noise power. The relationship between the input error probability and the SNR for binary signals in additive white Gaussian noise is well

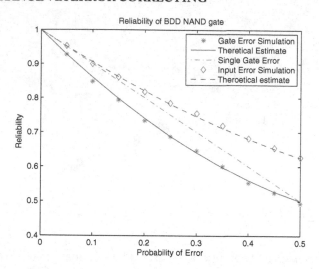

Figure 6.7: Reliability of a two-input NAND gate implemented as a BDD

known in communication [108]:

$$SNR = \frac{E}{2 \times \sigma^2},$$

where E is the energy of a binary pulse and σ is the signal deviation due to noise. It implies that the noise power is equal to $\sigma^2 = E/(2 \times SNR)$.

Fig. 6.8 shows the probability of error as a function of SNR. It follows from the graph, that the probability 0.1 of the gate (BDD output) error is due to an SNR of $2dB$. Once SNR increases to $12dB$, the probability of error is negligible, and, therefore, it is close to zero in the simulation. The Monte Carlo simulation is carried out by using a normally distributed random variable with zero mean and standard deviation.

Table 6.2 contains a list of SNR values, and the corresponding noise power, σ^2, used in calculations. The noise power is calculated based on the assumption that $V_{DD} = 0.3V$.

6.5   BDD MODEL WITH ERROR CORRECTION

Astola et al. [8] suggested combining the block error correcting codes and decision diagrams. Given a function f of k variables, its BDD includes k levels. If a switching error happens at a BDD node, it cannot be corrected. The corresponding error-correcting BDD can be designed as follows: a binary code (n, k) is constructed, and the function f is mapped into a function f', which is implemented using another BDD with n levels. Note that we are not interested here in mapping a code word C into an original message word A, but rather mapping it to the target value of the binary function $(C \rightarrow f(A))$. This means that no decoding is required, and the code words are mapped to the binary

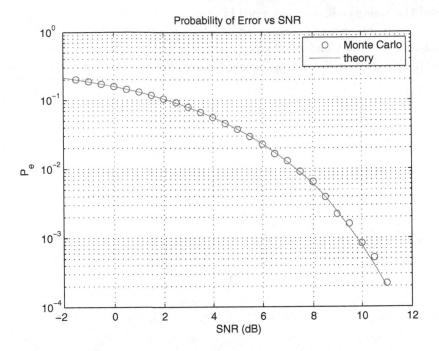

Figure 6.8: Probability of input error vs input signal SNR.

Table 6.2: Probability of gate error vs SNR and noise power given $V_{DD} = 0.3V$

SNR(dB)	Theoretical P_e	Noise power (σ^2)
-1.5	0.2	0.0318
0	0.1587	0.0225
2	0.1040	0.0142
3	0.0789	0.0113
4	0.0565	0.0090
5	0.0377	0.0071
7	0.0126	0.0045
9	0.0024	0.0028
10	0.0008	0.0022
12	0.0000	0.0014

0s or 1s. In this study, the Hamming code is used. The purpose of the study is to prove that the error-correcting BDD is able to withstand a single decision error (error in a BDD gate) due to the signal noise at the control input (input variable) of the node.

Shortened Hamming code.   A shortened Hamming code will be used in the case of an arbitrary number of inputs that cannot be expressed as $(2^m - m - 1)$. With shortened codes, a number of input vectors will have an undefined target, assigned with either 0 or 1. We choose to map these vectors to the binary value '0', and, optionally, to have an extra error indicator as shown in Fig. 6.9.

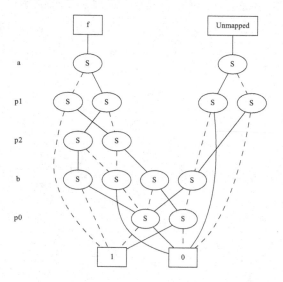

Figure 6.9: Error correcting NAND gate BDD with indicator for unmapped vector values (generated by the CUDD package [128]).

Buffer gate.   Consider the implementation of a buffer gate, using a one-level BDD (Fig. 6.10). This function of one variable corresponds to the $(3, 1)$ error-correcting code, in which 0 is encoded by 000, and one is encoded by 111. The decoder of such code represents a majority-vote function: it decodes the received codewords 000, 001, 010, and 100 as 0.

Error-correcting BDD for arbitrary functions.   This BDD can be the basis for generating the error correcting BDD of any elementary gate by replacing the terminal nodes with the values '0' and '1' and merging the diagram nodes accordingly. The generation of the parity bits is done by multiplying the message bits, A, by the parity matrix, P, using modulo-2 addition ($A \times P$ (mod 2)). Since this operation is exclusively binary, we can use a BDD to generate the parity bits from the message bits, instead of explicitly carrying out the matrix multiplication.

6.5.1   BDDS FOR SHORTENED CODES

Shortened codes   We saw that the Hamming codes are defined in terms of the parity bits such that the code is given as $(2^m - 1, 2^m - m - 1)$. For arbitrary message lengths, we can use the next higher value of $(2^m - m - 1)$ as the message length and encode all the extra bits as zeros. The position of

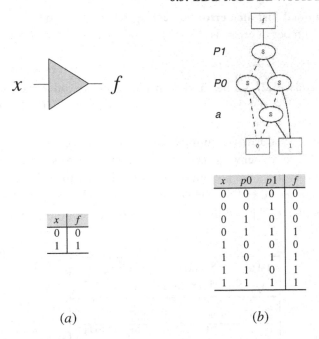

x	p0	p1	f
0	0	0	0
0	0	1	0
0	1	0	0
0	1	1	1
1	0	0	0
1	0	1	1
1	1	0	1
1	1	1	1

x	f
0	0
1	1

(a)   (b)

Figure 6.10: Error correcting BDD of a buffer gate: (a) the buffer gate, its error-free BDD and truth-table, and (b) the (3, 1) error correcting BDD and its corresponding truth-table (generated by the CUDD package [128]).

the extra zeros is arbitrary and we can thus truncate any rows of choice in the parity matrix. For example, if we want to define a Hamming code for a message of length $k = 2$, we start by choosing the nearest Hamming code; Hamming(7,4). Then we shorten the code by encoding each message as $[00a_1a_0]$ or $[a_1a_000]$ or $[0a_1a_00]$ or $[a_10a_00]$. This will effectively remove two rows from the Parity matrix making it $P_{2\times3}$. The generator matrix loses two rows and two columns as the identity matrix in the right-hand side loses two columns so that $G = [P_{2\times3}|I_2]$. This way we obtain the code words of the shortened Hamming code Hamming(5,2). This code is still capable of correcting only one error. The number of vectors that the code can correct is $2^k(1 + n)$. However, the number of vectors in the code space is 2^n. This means that a shortened Hamming code is not a perfect code as it does not map $2^n - 2^k(1 + n)$ input vectors. In the case of the non-perfect Hamming(5,2), the number of non-mapped vectors is $32 - 4 \times 6 = 8$. These non-mapped vectors can be dealt with by assigning them to an error indicator or by assigning them to one of the nearest code words according to the following definitions [85].

Definition 6.1   A maximum likelihood error correcting decoder is a decoder that given the received word r, selects the code word c which minimizes the Hamming distance $d_H(r, c)$.

Definition 6.2   **A bounded distance error correcting decoder** is a decoder that can select the correct codeword if the number of errors is $d_H(r, c) \leq t$. Otherwise, it signals decoder failure.

Shortened Hamming code (5,2).   The maximum likelihood decoder uses a *standard array* to achieve complete decoding. The standard array for the shortened code $H(5, 2)$ with parity matrix is shown in Table 6.3. The standard array is constructed by writing the code words (n-bit 2^k words) in the first row. Then rows 2 to 6 in the first column, we have the least weight error vectors ($weight = 1$). In rows 7 and 8 we put the next higher weight ($weight = 2$) error vectors. We choose these particular vectors because the values in them are not anywhere in the rows 1 to 6. A bounded distance error correcting decoder will only use rows 1 to 6 while a maximum likelihood decoder will use the whole standard array that covers all the 2^n possibilities. It can decode any single bit error or just two patterns of double bit errors.

Table 6.3: Standard decoding array for Hamming code (5,2)			
	00	01	10	11
row 1	00000	10101	11010	01111
row 2	00001	10100	11011	01110
row 3	00010	10110	11000	01101
row 4	00100	10001	11110	01011
row 5	01000	11101	10010	00111
row 6	10000	00101	01010	11111
row 7	00110	10011	11100	01001
row 8	01100	11001	10110	00011

For the shortened code (5,2), the parity generation diagram is shown in Fig. 6.11.

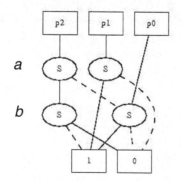

The BDD corresponds to the parity matrix
$$P = \begin{bmatrix} 1 & 1 & 0 \\ 1 & 0 & 1 \end{bmatrix}$$
The a and b are the bits of a 2-bit code word, and $p2$, $p1$, and $p0$ are the parity bits.

Figure 6.11: A BDD of a parity bit generator for the shortened Hamming code (5,2) (the BDD is generated by the CUDD package [128]).

The upper limit of complexity with block codes is doubling the number of the circuit elements while achieving the coding gain [107]. The increase in the BDD size has linear complexity. In the worst case, however, the unreduced BDD has a size that is exponential in the number of variables. Thus, augmenting the function variables by the parity bits will result in an exponential increase in complexity unless the function is decomposed.

Shortened Hamming code (7, 4) and error-correcting BDD for a 2-bit adder.   Consider the error-correcting BDD of a 2-bit adder implemented using the standard Hamming(7,4). The adder adds two binary words $A = a_1 a_0$ and $B = b_1 b_0$. The original shared diagram has 11 nodes, while the error correcting shared diagram has 61 nodes. Another alternative is to replace each node with four nodes representing the noise tolerant BDD of the buffer/inverter circuit. Therefore, the number of nodes in such a BDD is always $4n$. Note that the block coding theory implies that repeating each bit n times is inferior to block coding.

6.6   RELIABILITY OF ERROR-CORRECTING BDDS

In the reliability evaluation, we assume that error probability at any node is due to incorrect switching is p. The reliability of the BDD model is understood as the probability of reaching the correct value at the BDD output.

Buffer.   Consider an error correcting buffer gate shown in Fig. 6.10. If there are no errors at any of the decision nodes, then the reliability for this path is $(1 - p)^2$. If, however, there is a single error at any of the nodes, then there are two possible paths that have three nodes. The reliability for choosing one of those two paths is $2p(1 - p)^2$. Therefore, the total reliability of the BDD is:

$$R = (1 - p)^2 + 2p(1 - p)^2 = 1 - (3p^2 - 2p^3) = 1 - EP. \qquad (6.4)$$

Here, the error probability $EP = 3p^2 - 2p^3$. The same result can be reached using the properties of the error-correcting code itself. The reliability of the code is the probability that there is no error in any bit, except one (a single error occurs). Thus, the reliability of the code is estimated by the equation:

$$R = (1 - p)^3 + \binom{3}{1} p(1 - p)^2 = 1 - (3p^2 - 2p^3) \qquad (6.5)$$

which is the same result as the one obtained from the analysis of the BDD.

Figure 6.12 shows the simulation results of the reliability analysis for an error correcting BDD of the buffer/inverter gate, compared against the theoretical estimation using equation (6.5). Note that reference [9] provides a theoretical estimate of reliability that does not match the results from Monte Carlo simulation, according to the above figure. For gates other than the buffer/inverter, the reliability becomes a function of the truth table, and not the error correction capability of the code. This is because we are not interested in restoring the inputs of the gate function, but rather in getting the correct output. Thus, while calculating the gate reliability, we assume the possibility that any change of the input signal does not result in any change in the output signal.

NAND and XOR gates. Consider an array for the shortened Hamming code (5,2) (Table 6.3). Given a NAND gate, the first three columns should correspond to the output value '1' and the last column - to the output value '0'. The error is defined as a change in the output value from 0 (1) to 1 (0). To go from a value in any column to an opposite one in any other column, we need to change at least two bit positions, as the minimum Hamming distance of this code is 3. The reliability of the error correcting AND/NAND gate includes three components: the probability of staying within the same column (for the first six rows); the probability of one of the first three columns (also within the first six rows) changing to another one; and the probability of the value in the fourth column changing to any of the values in the 7th and 8th rows, as these rows are mapped to the value 0:

$$R = R_{EC} + \frac{1}{2}(R_{00\leftrightarrow01} + R_{00\leftrightarrow10}) + \frac{1}{2}R_{10\leftrightarrow01} + \frac{1}{4}R_{unmapped} \tag{6.6}$$

$$= (1-p)^5 + 5p(1-p)^4 + \frac{3}{4}(3p^3(1-p)^2 + 6p^2(1-p)^3 + 3p^4(1-p)) + \frac{1}{4}(p^2(1-p)^3).$$

Equation (6.6) does not take into account the structure of the diagram itself, but uses the properties of error correction coding in conjunction with the truth table of the gate. The reliability of the XOR gate is derived in a similar way.

Figure 6.13 illustrates the theoretical value of the reliability of an AND/NAND gate, using equation (6.6), and the result of a Monte Carlo simulation of the error-correcting BDD based on Hamming code (MC Hamming). Random errors were applied independently to each signal at all BDD levels. It also shows the effect of replacing each node of the BDD for the NAND gate with an error-correcting BDD node based on the triple-modular redundancy models of a buffer/inverter (denoted in figure as MC TMR). This resulted in a six-level $4 \times 2 = 8$-node BDD, as opposed to a five-level 10-node error-correcting BDD for the NAND gate. The Monte Carlo simulation used independent noise at each of the six levels. The reliability of the error-correcting BDD for an XOR/XNOR gate is shown in Figure 6.14.

2-bit adder. Given a 2-bit adder, its BDD has three outputs and 11 nodes. Its triple-modular redundancy (TMR) model has 44 nodes, and the model based on the Hamming code has 61 nodes. The simulation for reliability estimation of these models for each of the individual outputs s_2, s_1, s_0 of the adder is shown in Figure 6.15.

In simulations, a high-level multiplexer model of a node is considered. This model is used to analyze the performance of a 2-bit adder and assess the bit-error rate averaged over the three outputs of the adder as the SNR is degraded. A fragment of the simulation results at $SNR = 9dB$ is shown in Fig. 6.16, and results for other SNR values are given in Table 6.4. The simulation demonstrated that with error correction enables almost an order of magnitude improvement in the device noise-tolerance.

To model circuit delays in a larger circuit simplified bidirectional hysteresis switch circuit model was used. This model allows for reduction of the simulation time. Each BDD node is represented using a dual transmission gate consisting of two pairs of pass-transistors. An LP 16-nm predictive

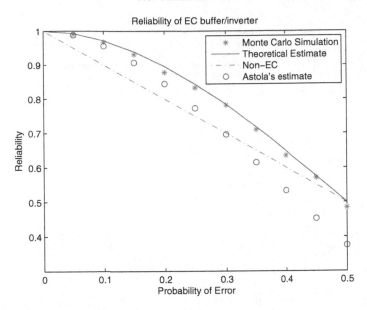

Figure 6.12: Reliability of the error correcting BDD for the buffer/inverter

Table 6.4: Noise tolerance in error-correcting (EC) BDD models of a two-bit adder, measured in BERs with uncorrelated noise added at all four inputs for various SNR levels

SNR	Conventional design	EC BDD design
3	0.4126	0.1650
5	0.3236	0.0871
7	0.2080	0.0283
9	0.0933	0.0052
10	0.0510	0.0016
12	0.0107	0

CMOS technology model from [109] was used for simulation in Spice. The noise was added to the control signal, using the noise mean at zero and the standard deviation. The gate voltage in the simulations was set above threshold, at V_{DD}. Fig. 6.17 shows how the transistor-level circuit handles the large superimposed noise at all switching levels of the BDD.

Planar BDD arrays.   In order to realize a circuit such as error-correcting BDD models of an adder on a GaAs hexagonal BDD array [58, 60], it is mandatory to derive a BDD in the forms of a planar graph.

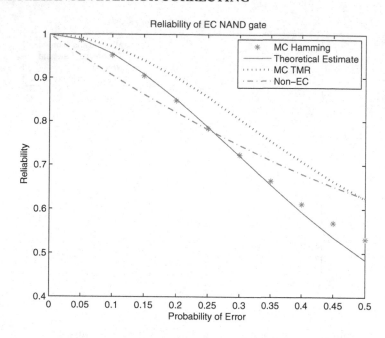

Figure 6.13: Reliability of the error correcting BDD for the AND/NAND gate

This can be achieved by using an algorithm proposed in [17]. The transformation of an arbitrary BDD into a planar BDD involves the following manipulations: (*a*) inserting dummy nodes, (*b*) swapping two nodes, (*c*) duplicating nodes, and (*d*) duplicating edges. First a reduced ordered BDD is generated. Next, each BDD level is examined, and the nodes of this level are arranged according to the position of their parent nodes (from the upper level). If a node has more than one parent, and the parents are not adjacent within their level, then this node is duplicated. The process is repeated until all nodes are processed in a single pass. The generated diagram is neither canonical nor optimal. To reduce it to an optimal planar diagram, the possibility for the node sharing is examined. To maximize node sharing, the parent nodes are shuffled. This shuffling is repeated until the optimal number of the shared nodes is achieved. Without parent node shuffling, the complexity and the processing time is linear in the number of nodes, and the whole diagram is processed in one pass. Two examples of a planarized error-correcting BDDs: one for a NAND gate and the other one for the output s_2 of a 2-bit adder are shown in Fig. 6.18 and Fig. 6.19, respectively.

The structure of the planar BDDs satisfies the requirements of implementing on a hexagonal GaAs-based nanowire network [58, 161].

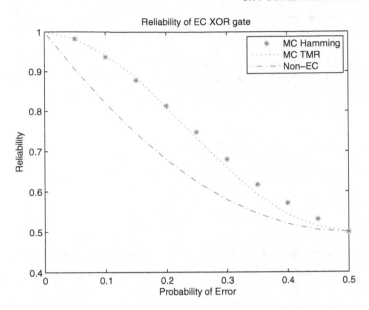

Figure 6.14: Reliability of the error correcting BDD for the XOR/XNOR gate.

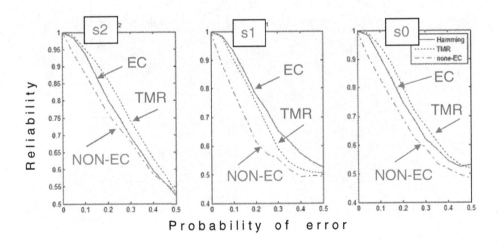

Figure 6.15: Comparison of the reliability of a 2-bit adder design using the error-correcting (EC) BDD, the triple-modular redundancy (TMR), and CMOS gate-level adder design (denoted by None-EC).

6.7   SUMMARY AND DISCUSSION

This chapter provides a brief introduction to error-correcting BDD-based computational data structure, based on the approach by Astola's et al. [8, 9]. It formalizes the design of such BDDs by adding

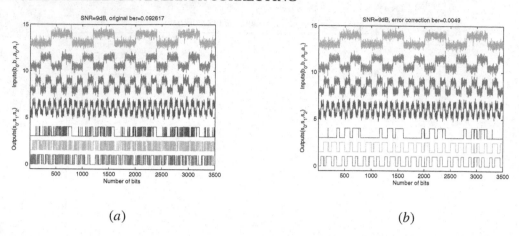

(a) (b)

Figure 6.16: (a) Simulation of the two-bit adder without error correction at $SNR = 9dB$. (b) Simulation of the adder with error correction. BER values are averaged for all three output bits.

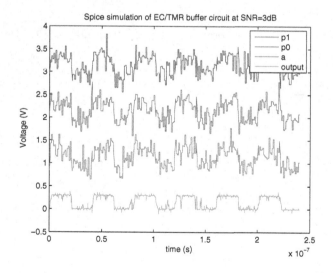

Figure 6.17: Spice simulation results for the error-correcting BDD model of a buffer, using various random noise applied to the control signal at each BDD level.

the procedure for generating the parity-code BDDs. It also extends this approach toward obtaining the planar error-correcting BDDs. In particular, the following results are reported:

1. This approach to noise-tolerant logic gate design is different from the probabilistic models based on maximum likelihood computing paradigm, such as Boltzmann machine and the

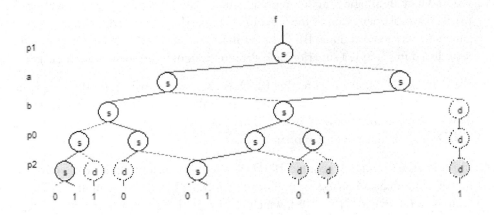

Figure 6.18: A planar error-correcting BDD for a NAND gate.

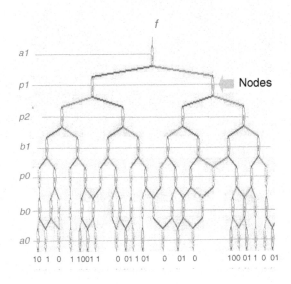

Figure 6.19: A planar error-correcting EC-BDD for the output $s2$ of a 2-bit adder.

MRF model. The error-correcting BDD-based model explores the domain of error-tolerance via adding hardware redundancy. Because of this redundancy, the true value of a logic function is decoded despite error in the BDD nodes (interpreted as incorrect switching in the nodes due to errors in control signals, corresponding to the input variables of the logic function of a gate).

2. Redundancy in a logic function representation plays a useful role in combating the noise in the BDD implementation of the function. Comparison of the error-correcting BDDs of logic gates and their conventional BDD-based analogs confirms that the error-correcting approach introduced in [8, 9] is a superior candidate for designing noise-tolerant logic gates.

3. Planarization of the error-correcting BDDs is a necessary attribute for their implementation on switch-based (2-to-1 multiplexer-based) nanoscale devices, such as on the BDD-based hexagonal nano-wire BDD arrays [38, 60]. This chapter introduces an approach for planarization of BDDs via a linear-cost redundancy.

4. In future, the Hamming codes in BDD should be replaced with more effective codes, for instance, random error-correcting cyclic codes, or codes based on maximum likelihood principle, such as iterative codes [71, 108], which can be implemented using a special type of BDDs with feedback called cyclic BDD (see Chapter 4).

CHAPTER 7

Conclusion and future work

Murphy' reminder

80% of the final exam will be based on the
one lecture you missed

This book includes selected lectures adopted for graduated courses in electrical and computing engineering programs at the University of Calgary (Canada) and Hokkaido University (Japan). These lectures cover

(*a*) **Design** of noise-resilient logic gates and circuits, and

(*b*) **Implementation** of such devices in nanoscale, and represent the introduction in the art of modern noise-resilient computation.

Computation in the presence of noise, also called probabilistic computation, has been a classical topic since the beginning of the computer era [86, 88, 122, 143]. Note that the basic relationship between Boolean algebra and the rules for manipulation of probabilities events were well understood and exploited by George Boole in 1854. T. Hailperin in his book "Boole's Logic and Probability" wrote: "Never clearly understood, and considered anyhow to be wrong, Boole's ideas on probability were simply by-passed by the history of the subject, which developed along other lines" [39].

It is paradoxical that now again, same as at the dawn of computers, **reliable computation using unreliable elements** has become a core topic of nanoscale digital design. However, the problem must be reformulated, in the light of the physical and chemical phenomena at nanoscale, as emphasized in Chapter 2 of this book. When the supply voltage becomes comparable to thermal noise levels in ultra-low power designs, devices start to behave probabilistically giving an incorrect output with some nonzero probability. At the nanoscale, devices are more stochastic in operation, and quantum effects become the rule rather than the exception. It is unlikely that existing computational models are an optimal mapping to these new devices and technologies, and this is the motivation for new computing paradigms. The International Technology Roadmap for Semiconductors predicts that **thermal noise will cause devices to fail catastrophically** during normal operation without supply voltage scaling [46]. That is, in the subthreshold region device operates below threshold voltage, and it is sometimes approximated as being "turned off."

A taxonomy of probabilistic computing using such a new formulation has not been developed yet, except for particular solutions. In this book, we have introduced several probabilistic models for noise-resilient design of logic gates and networks. We have also formulated and indicated the open problems in this area. For example, Chapter 6 on error correcting BDDs is as invitation to develop more efficient noise-resilient techniques using iterative codes and cyclic, or loopy, structures. We summarize these insights as follows.

1. Probabilistic techniques become an inherent part of noise-resilient computing technology. The theoretical platform of this book includes:

(*a*) The fundamentals of computations in the presence of noise, which were initially developed by Shannon and Moore [86, 122], von Neumann [143], and Winograd and Cowan [146].

(*b*) The advanced probabilistic models for computing under uncertainty, such as Bayesian networks [52, 104, 105] the MRF model [13, 23, 35], and other techniques [26, 66, 68, 120].

(*c*) The fundamentals of neuromorphic computations [82]; specifically, Hopfield networks and Boltzmann machines [41, 44].

In 1956, von Neumann showed that **reliable computation can be achieved by circuits with noisy gates** [143]. Von Neumann's techniques resulted in unacceptable hardware overhead, however, it is still discussed as a possible solution for predictable technologies [37, 112]. It was well documented that logic gates and networks propagate error not equally, regardless of specific configuration. For example, the deeper a network's depth (number of levels), the more fault-tolerant the circuit tends to be given a fixed number of faults [88, 103]. It is known that feedback networks have the property of alleviation of noise [40]. Today, it is well understood that technological implementation introduces so-called **dynamic errors**, which are difficult to detect and correct using traditional methods and tools.

We distinguish probabilistic models for **software and hardware** implementation, focusing on the latest for logic gates (simplest switching functions). The essence of the advanced hardware-centric techniques for computation under uncertainty lies in the following. Direct manipulation with probabilities (real numbers) are replaced by **indirect computation** using special mechanism of **approximation probability distribution** by arithmetic sum of minterms, taken from compatibility truth table, and **maximum likelihood principle**, applied to these minterms. This iterative process is implemented using various types of circuits with feedbacks, such as PLA with feedbacks, cyclic BDDs, and sets of completely and bidirectionally connected threshold gates.

Neuromorphic networks are viewed to be biologically inspired computing networks and memory storage; they are supposed to inherit the following properties of the brain: robustness and fault tolerance, flexibility, processing of fuzzy, probabilistic, noisy, or inconsistent information, and massively parallel and distributed processing [82]. The familiar examples of these models are the Hopfield network and Boltzmann machine. In 1982, the American physicist John Hopfield proposed an asynchronous network of threshold cells known as Hopfield network, or Hopfield computing paradigm, which is based on the concept of energy minimization in a stochastic system.

2. New taxonomy of noise-resilient techniques is needed. Being established as an intrinsic part of computer design, probabilistic techniques might be revised, and new taxonomy is needed in ultra-deep submicron and predictable technologies. The emergency of the problem is caused by scaling of feature sizes and voltages in modern semiconductor process technologies. The ultra-low power computing, which is achieved by lowering the supply voltage of digital circuits into near threshold, or the subthreshold region, must be addressed as well. Various noise-resilient techniques have been proposed recently, in particular [78, 102, 133, 141]. This book contributes to noise-resilient design of logic gates which are the fundamental building blocks for various technologies. We focus on the models, which exploit the **local computation principle** (the best idea to deal with joint probability distribution), and are well suited for hardware implementation (indirect manipulation with real numbers, that is, probabilities). From general point of view, noise-resilient computation is an essence an **information-encoding problem**, consisting in iterative improving the probability of correct output **similar to the iterative decoding**. Note that analog devices often provide good noise-resilient properties. In **analog probabilistic models**, every signal is continuous, taking a value between logical 0 or 1. Hence, the probabilistic models are able to propagate probabilities and to extend models by exploring, for example, fuzzy logic.

3. Switch-based data structures, such as decision diagrams, are attractive for nanoscale technology. Among various data structures of switching functions, graphical forms such as decision trees and diagrams are mainly used in this book (it is assumed that graduated students have prerequisite knowledge, summarized, in particular, in [153]). However, classical decision diagrams implement a **feed-forward** type of computation, and, therefore, cannot be used for computation with feedback (Fig. 7.1a). An iterative (recurrent) probabilistic model distinguishes itself from the feed-forward models as it has at least one feedback loop. We have shown that iterative probabilistic models can be implemented by so-called **cyclic** decision trees and diagrams. That is, we proposed an extension of conventional decision diagrams, called the cyclic decision diagrams (Fig. 7.1b) [158]. They are switch-based computing data structures with feedback. Note that in deep submicron technology, the energy for simplest switching operation $0 \rightarrow 1$ on a single bit is nearly $10^4 - 10^5$ kT, compared to 20 kT of an elementary noise-resilient operation in a biological cell.

In probabilistic models, switch-based data structures are used indirectly, because the structures operate on single bit, while the probability is a real value. Decision diagrams play the role of **"topological frames"** in the probabilistic modeling; that is, the initial and current probabilities are assigned to the diagram edges. It is assumed that computation with real valued numbers is performed separately, according to the diagram topology and the node function. Thus, a decision diagram on Fig. 7.1, left, is the frame for **bottom-up computation** of probabilities. Let us assume that probabilities are assigned to edges of the cyclic decision diagram shown in Fig. 7.1, right. The most probabilistic paths (minterms) can be **reinforced** using iterative computing.

4. Hardware-centric noise-resilient computing paradigms are needed for designing logic networks. The focus of this book is probabilistic models of logic gates, which are considered as

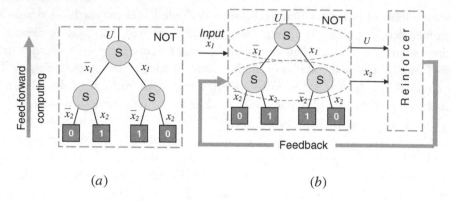

(a) (b)

Figure 7.1: Two types of decision trees (diagrams) using compatibility truth table of the NOT function (gate): (a) classical feed-forward principle of computation, and (b) computation with a feedback.

candidates for hardware implementation. The main problem is how to replace costly operations on real numbers by operations on bits. A computing paradigm based on **relaxation** provides such a possibility (Fig. 7.2, first two models). This model utilizes **undirected** graphical data structures and **local principle** of computation, but functions of nodes are different. However, locality in the Hopfield (Boltzmann) model is specified by complete interconnections of nodes. In MRF, a special procedure is needed for finding a set of local nodes. Also, the implementation of these models are different. Both models provide indirected probabilistic computation. Specifically, **a logic function is incorporated into the probabilistic model** using its compatibility truth table, approximation of the probability distribution by the sum of minterm probabilities, and application of the maximum likelihood principle via reinforcing of the most probable minterm(s). Thus, this indirected probabilistic computation can be implemented by a **bit operation**, and is acceptable for hardware implementation.

The last two models exploit the principle of belief propagation. In a Bayesian network, belief is propagated and is represented by real values. For computation, adder and multiplier of real numbers are required. This model can be used for analysis, but it is not acceptable for hardware implementation of logic gates. As mentioned above, the decision diagrams are the "topological frame" for probabilities, assigned to the diagram edges.

5. Challenges.   The proposed collection of selected lectures should be considered as an introduction to the area of noise-resilient computing. There exist other diverse techniques and innovative technologies in related areas, which can be adopted for designing noise-tolerant logic gates, but were not included in the book. In particular, Pearl's belief propagation principle in the Bayesian models [104, 105] is a key to understanding the probabilistic techniques, such as iterative codes in communication. Also, turbo decoding, introduced in 1993 by Berrou et al. [11], is the most exciting and potentially important development not only in the coding theory, but also in the design of noise-resilient computing devices (the attempts to apply the turbo decoding principle to logic

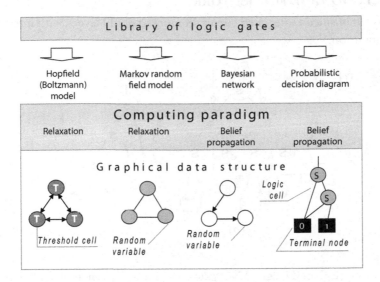

Figure 7.2: Comparison of noise-resilient logic gate models, according to the criteria of computing organization and topology of computational data structures.

network design include [12, 12, 97, 110]). Note that the method of turbo decoding is equivalent to the loopy belief propagation, if an appropriate graphical model is used[31, 81].

6. Toward spatial noise-resilient computing devices. The brain has a 3D interconnect technology that enables a fan-in and fan-out of nearly 6,000 connections per neuron, compared with about 10 in the conventional CMOS logic networks. The brain consists of about 22 billion neurons with power dissipation about 15W. This is due to efficient 3D technology integration, expedient 3D design of neuron networks, and highly energy-efficient and noise-resilient signal processing. Electronics, inspired by neurobiology, was termed by Mead "neuromorphic electronics" [82].

It has been proved in a number of papers that 3D topologies of computer elements are optimal, physically realistic models for scalable computers [30]. In contemporary logic network design, the use of the third dimension is motivated by decreasing interconnection topology. The third dimension is thought of as **layering**. For example, in chips, networks are typically assembled from about ten layers [36]. A 3D stacking and integration can provide significant system advantages: they alleviate the interconnect I/O bandwidth and latency bottlenecks, by leveraging the third axis to decrease communication distances and exchange information between blocks. For some devices, such as cell phones and memory, 3D integration (also referred to as system-in-package) is already in use. A typical problem is that the area footprint of the through-silicon vias are comparable to that of gates. For example, at the 45 nm technology, the area of a TSV is $10\mu m \times 10\mu m = 50$ gates [62].

Our approach to embedding the BDDs into 3D space is illustrated in Fig. 7.3 using the 3-input AND gate, $f = x_1x_2x_3$ [125]. The result, a 3D hypercube, has a truly 3D structure, instead of the 3D layout of silicon integrated circuits composed of 2D layers, with interconnections forming the third dimension.

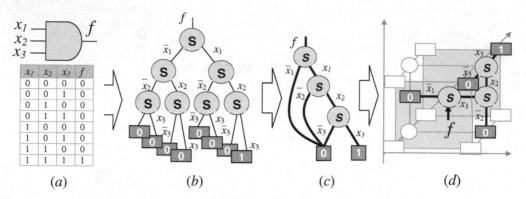

(a)   (b)   (c)   (d)

Figure 7.3: A noise-free 3D logic AND gate design using decision diagram data structure: (a) truth table, (b) decision tree, (c) decision diagram, and (d) a 3D decision diagram as the result of embedding 2D decision into hypercube. The last question in final examination is as follows: "Given a probabilistic model, provide its computation abilities in 3D dimension."

Various extensions of this technique for probabilistic models are straightforward, including 3D cyclic BDDs. Note that in the proposed spatial model, the third dimension is not considered as arising merely from layering, such as in TSV-based 3D integration technology, but as a **dimension for achieving the desired functional logical properties** [152, 157].

Bibliography

[1] S. Adachi and K. Oe, Chemical etching characteristics of (001) GaAs, *J. Electrochem. Soc,* vol. 130, issue 12, pp. 2427–2435, 1983. DOI: 10.1149/1.2119608 25

[2] V. Agrawal, An information-theoretic approach to digital fault testing, *IEEE Trans. Comput.,* vol. 30, no.8, pp.582–587, 1981. DOI: 10.1109/TC.1981.1675843 6, 7, 35, 41, 94

[3] A. Agrawal, K. Kang, S. Bhunia, J. Gallagher, and K. Roy, Device-aware yield-centric dual-V_t design under parameter variations in nanoscale technologies, *IEEE Trans. VLSI,* vol. 15, no. 6, pp.660–671, 2007. DOI: 10.1109/TVLSI.2007.898683 2

[4] S. B. Akers, A rectangular logic array, *IEEE Trans.,* 21:848–857, Aug. 1972. DOI: 10.1109/TC.1972.5009040 27

[5] L. P. Alarcon, T.-T. Liu, M. D. Pierson, and J. M. Rabaey, Exploring very low-energy logic: A case study, *J. Low Power Electronics,* vol. 3, pp.223–233, 2007. DOI: 10.1166/jolpe.2007.136 2

[6] C. Arnd, *Information Measures: Information and Its Description in Science and Engineering.* Springer, Heidelberg, 2001. 42

[7] R. B. Ash, *Information Theory,* Wiley, New York, 1967. 42, 94

[8] H. Astola, S. Stanković, and J. T. Astola, Error-correcting decision diagrams, *Proc. 3rd Workshop Information Theoretic Methods in Science and Engineering,* Tampere, Finland, Aug., 2010. DOI: 10.1109/ISMVL.2011.33 xvi, 87, 98, 107, 110

[9] H. Astola, S. Stanković, and J. T. Astola, Error-correcting decision diagrams for multiple-valued functions, *Proc. 41st IEEE Int. Symp. Multiple-Valued Logic,* 2011. DOI: 10.1109/ISMVL.2011.33 103, 107, 110

[10] R. I. Bahar, J. Chen, and J. Mundy, A probabilistic-based design for nanoscale computation. In: S. Shukla, and R.I. Bahar, Eds. *Nano, Quantum and Molecular Computing: Implications to High Level Design and Validation,* Kluwer, 2004. 11, 49, 52, 53, 70

[11] C. Berrou, A. Glavieux, and P. Thitimajshima, Near Shannon limit error-correcting coding and decoding turbo-codes, *Proc. IEEE Int. Conf. Commun.,* Geneva, pp. 1064–1070, May, 1993. DOI: 10.1109/ICC.1993.397441 1, 5, 17, 45, 88, 114

[12] C. Berrou and V. Gripon, Coded Hopfield networks, *Proc. 6th Int. Symp. Turbo Codes and Iterative Information Processing,* 2010. DOI: 10.1109/ISTC.2010.5613860 88, 115

[13] J. E. Besag, Spatial interaction and the statistical analysis of lattice systems, *J. Roy. Stat. Soc.*, series B, vol. 36, no. 3, pp.192–236, 1974. 51, 112

[14] R. C. Born and A. K. Scidmore, Transformation of switching functions to completely symmetric switching functions, *IEEE Trans. Comput.*, vol. 17, no 6, pp. 596–599, 1968. DOI: 10.1109/TC.1968.226926 28

[15] B. D. Brown and H. C. Card, Stochastic neural computing I: Computational elements, *IEEE Trans. Comput.*, vol. 50, no. 9, pp.891–905, 2001. DOI: 10.1109/12.954505 6, 16

[16] R. E. Bryant, Graph-based algorithms for Boolean function manipulation, *IEEE Trans. Comp.*, vol. 35, no. 6, pp.677–691, 1986. DOI: 10.1109/TC.1986.1676819 36

[17] A. Cao and C. K. Koh, Non-crossing OBDDs for mapping to regular circuit structures, *Proc. 21st Int. Conf. Comp. Design*, pp.338–343, 2003. DOI: 10.1109/ICCD.2003.1240916 106

[18] S. T. Chakradhar, V. D. Agrawal, and M. L. Bushnell, *Neural Models and Algorithms for Digital Testing*, Kluwer, Dordrecht, 1991. DOI: 10.1007/978-1-4615-3958-2 17, 81

[19] V. Cheushev, C. Moraga, V. Shmerko, S. Yanushkevich, and J. Kolodziejszyk, Information theory method for network synthesis, *Proc. IEEE 31th Int. Symp. Multiple-Valued Logic*, pp.201–206, 2001. DOI: 10.1109/ISMVL.2001.924573 7, 35, 41, 97

[20] V.A. Cheushev, S. N. Yanushkevich, C. Moraga, and V. P. Shmerko, Flexibility in logic design, An approach based on information theory methods. *Research Report*, Forschungsbericht 741, University of Dortmund, Germany, 2000. 7, 41, 97

[21] N. Clement, K. Nishiguchi, A. Fujiwara, and D. Vuillaume, One-by-one trap activation in silicon nanowire transistors, *Nature Communications*, vol. 1, art. 1092, 2010. DOI: 10.1038/ncomms1092 31

[22] R. H. Dennard, F. H. Gaensslen, H.-N. Yu, V. L. Rideout, E. Bassous, and A. R. LeBlanc, Design of ion-implanted MOSFET's with very small physical dimensions, *IEEE J. of Solid-State Circuits*, vol. SC-9, no.5, pp. 256–268, 1974. DOI: 10.1109/JPROC.1999.752522 20

[23] H. Derin and P. A. Kelly, Discrete-index Markov-type random process, *Proc. IEEE*, vol. 77, no. 10, pp. 1485–1510, 1989. DOI: 10.1109/5.40665 49, 112

[24] K. Dick, K. Deppert, M. W. Larsson, T. Martensson, W. Seifert, L. R. Wallenberg, and L. Samuelson, Synthesis of branched 'nanotrees' by controlled seeding of multiple branching events, *Nature Material*, vol.3 , no. 6, pp. 380–383, 2004. DOI: 10.1038/nmat1133 25

[25] M. Drapeau, V. Wiaux, E. Hendrickx, S. Verhaegen, and T. Machida, Double patterning design split implementation and validation for the 32nm node, *Proc. SPIE*, no. 6521 p.652109, 2007. DOI: 10.1117/12.712139 22

[26] D. Dubois and H. Prade, *Possibility Theory: An Approach to Computerized Processing of Uncertainty*, Plenum, New York, 1988. 37, 112

[27] R. O. Duda, P.E. Hart, and D.G. Stork, *Pattern Classification*, 2nd ed., Wiley, 2001. 2, 8, 44, 81

[28] T. J. Dysart and P. M. Kogge, Analyzing the inherent reliability of moderately sized magnetic and electrostatic QCA circuits via probabilistic transfer matrices, *IEEE Trans. VLSI Systems*, vol. 17, no. 4, pp.507–516, 2009. DOI: 10.1109/TVLSI.2008.2008092 5

[29] E. Fredkin and T. Toffoli, Coservative logic, *Int. J. Theoretical Physics*, vol. 21, Nos. 3/4, pp. 219–253, 1982. DOI: 10.1007/BF01857727 15

[30] M. P. Frank and T. F. Jr. Knight, Ultimate theoretical models of nanocomputers, *Nanotechnology*, vol. 9, pp.162–176, 1998. DOI: 10.1088/0957-4484/9/3/005 115

[31] B. J. Frey and N. Jojic, A comparison of algorithms for inference and learning in probabilistic graphical models, *IEEE Trans. Pattern Analysis and Machine Intelligence*, vol. 27, no. 9, pp.1392–1416, 2005. DOI: 10.1109/TPAMI.2005.169 17, 49, 115

[32] T. Fukui, S. Ando, Y. Tokura, and T. Toriyama, GaAs tetrahedral quantum dot structure fabricated using selective area metalorganic chemical vapor deposition, *Appl. Phys. Lett.*, vol. 58, no.18, pp. 2018–2020, 1991. DOI: 10.1063/1.105026 23

[33] T. Fukushi, T. Muranaka, and H. Hasegawa, Atomic hydrogen-assisted selective MBE growth of hexagonal InGaAs ridge quantum wire networks having a high density of Giganodes/cm2, *Proc. Int. Conf. Indium Phosphide and Related Materials (IPRM)* pp.315–318, 2003. DOI: 10.1109/ICIPRM.2001.929196

[34] B. R. Gaines, Stochastic computing systems, in "Advances in Information Systems Science," J. T. Tou, Ed., Plenum, New York, vol. 2, chap. 2, pp.37–172, 1969. DOI: 10.1007/978-1-4615-9050-7 6, 16

[35] S. Geman and D. Geman, Stochastic relaxation, Gibbs distributions and the Bayesian restoration of images, *IEEE Trans. Pattern Analysis and Machine Intelligence*, vol. 6, pp.721–741, 1984. DOI: 10.1109/TPAMI.1984.4767596 8, 11, 41, 49, 51, 112

[36] Global Industry Analysts, Inc. (2010). 3D Chips (3D IC): A Global Market Report [Online]. Available: http://www.prweb.com/releases/3D chips/3D IC/prweb4400904.htm 115

[37] J. Han and P. Jonker, A system archtecture solution for unreliable nanoelectronic devices, *IEEE Trans. Nanotechnology*, vol. 1, no. 4, pp.201–208, 2002. DOI: 10.1109/TNANO.2002.807393 6, 112

[38] H. Hasegawa, S. Kasai, and T. Sato, Hexagonal binary decision diagram quantum circuit approach for ultra-low power III-V quantum LSIs, *IEICE Trans. Electron.*, Vol. Es7-C. no.11, pp.1757–1768, 2004. xvi, 60, 70, 110

[39] T. Hailperin, *Boole's Logic and Probability*, 2nd ed., Elsevier, Amsterdam, 1986. 111

[40] S. Haykin, *Neural Networks and Leaning Machines*, 3rd edition, Pearson Education, Upper Saddle River, NY, 2009. 2, 8, 13, 14, 15, 16, 41, 43, 50, 52, 54, 75, 77, 81, 112

[41] G. E. Hinton and T. J. Sejnowski, Optimal perceptual inference, *Proc. IEEE Conf. Computer Vision and Pattern Recognition,* vol. 352, pp.448–453, Washington, DC, 1983. 112

[42] G. E. Hinton, Deterministic Boltzmann machine learning performs steepest descent in weight-space, *Neural Computation*, vol. 1, pp.143–150, 1989. DOI: 10.1162/neco.1989.1.1.143 8, 17, 80

[43] F. N. Hooge, $1/f$ noise sources, *IEEE Tran. Electron Devices*, vol. 41, no. 11, pp.1926–1935, 1994. DOI: 10.1109/16.333808 32

[44] J. J. Hopfield, Neural networks and physical systems with emergent collective computational abilities, *Proc. National Academy of Sciences*, USA, vol. 79, pp.2554–2558, 1982. DOI: 10.1073/pnas.79.8.2554 8, 15, 17, 76, 112

[45] J. J. Hopfield, Neurons with graded response have collective computational properties like those of two state neurons, Proc. National Academy of Sciences, vol.81, no.10, pp.3088–3092, 1984. DOI: 10.1073/pnas.81.10.3088 76

[46] http://www.itrs.net 22, 111

[47] D. A. Huffman, Combinational circuits with feedback. In: Mukhopadhyay, A. (Ed.), *Recent Developments in Switching Theory*. Academic Press, 1971, pp. 27–55. 15

[48] S. Iijima, Helical microtubules of graphitic carbon, *Nature*, no. 354, pp.56–58, 1991. DOI: 10.1038/354056a0 22

[49] ITRS, http://www.itrs.net/reports.html, 2012. 63, 69

[50] C. L. Janer, J. M. Quero, J. G. Ortega, and L. G. Franquelo, Fully parallel stochastic computation architecture, *IEEE Trans. Signal Processing*, vol. 44, no. 8, pp.2110–2117, 1996. DOI: 10.1109/78.533736 5, 6, 16

[51] S. Jankowski, A. Lozowski, and J. M. Zurada, Complex-valued multistate neural associative memory, *IEEE Trans. Neural Networks*, vol. 7, no. 6, pp.1491–1496, 1996. DOI: 10.1109/72.548176 84

[52] F. V. Jensen, *Bayesian Networks and Decision Graphs*, Springer, 2001. 1, 8, 17, 35, 112

[53] R. Jia, H. Hasegawa, N. Shiozaki, and S. Kasai, Device interference in GaAs quantum wire transistors and its suppression by surface passivation using Si interface control layer, *J. Vac. Sci. & Technol. B* , vol. 24, no.4, pp. 2060–2068, 2006. DOI: 10.1116/1.2216720 33

[54] H. Jung and M. Pedram, Resilient dynamic power management under uncertainty, *Proc. Int. Conf. Design Automation and Test in Europe (DATE)*, pp.224–229, 2008. DOI: 10.1145/1403375.1403430 2, 43

[55] Z. Kamar and K. Nepal, Noise margin-optimized ternary CMOS SRAM delay and sizing characteristics, *Proc. IEEE Int. Midwest Symp. Circuits Syst.*, pp.801–804, 2010. DOI: 10.1109/MWSCAS.2010.5548690 65, 66, 67, 68

[56] D. Kasahara, Leonard, K. Pond, and P. M. Petroff, Critical layer thickness for self-assembled InAs islands on GaAs, *Phys. Rev, B*, vol. 50, no.16, pp. 11687–11692, 1994. DOI: 10.1103/PhysRevB.50.11687 22

[57] J. Kasahara, K. Kajiwara, and T. Yamada, GaAs whiskers grown by a thermal decomposition method, *J. Crystal Growth*, vol. 38, no.1, pp.23–28, 1977. DOI: 10.1016/0022-0248(77)90368-2 22

[58] S. Kasai and H. Hasegawa, A single electron BDD quantum logic circuit based on Schottky wrap gate control of a GaAs nanowire hexagon, *IEEE Electron Device Letters*, vol.23, no. 8, pp.446–448, 2002. DOI: 10.1109/LED.2002.801291 24, 54, 55, 97, 105, 106

[59] S. Kasai, T. Hashizume, and H. Hasegawa, Electron beam induced current characterization of novel GaAs quantum nanostructures based on potential modulation of two-dimensional electron gas by Schottky in-plane gates, *Jpn. J. Appl. Phys.*, no. 35, pp.6652–6658, 1996. DOI: 10.1143/JJAP.35.6652 26

[60] S. Kasai, T. Nakamura, Y. Shiratori, and T. Tamura, Schottky wrap gate control of semiconductor nanowire networks for novel quantum nanodevice-integrated logic circuits utilizing BDD architecture, *J. Computational and Theoretical Nanoscience*, vol.4, no. 6, pp. 1120–1132, 2007. DOI: 10.1166/jctn.2007.004 24, 97, 105, 110

[61] W. H. Kautz, The necessity of closed loops in minimal combinational circuits, *IEEE Trans. Comput.*, no. 2, pp.162–164, 1970. DOI: 10.1109/T-C.1970.222884 15

[62] D. H. Kim, S. Mukhopadhyay, and S. K. Lim, Through-silicon-via aware interconnect prediction and optimization for 3D stacked ICs, *Proc. Int. Workshop Syst.-Level Interconn. Pred.*, pp. 85–92, 2009. DOI: 10.1145/1572471.1572486 115

[63] M. J. Kirton and M. J. Uren, Noise in solid-state microstructures: A new perspective on individual defects, interface states and low-frequency $(1/f)$ noise, *Advances in Physics*, vol. 38, no.4, pp. 367–468, 1989. DOI: 10.1080/00018738900101122 31

122 BIBLIOGRAPHY

[64] P. Korkmaz, B. E. S. Akgul, and K. V. Palem, Energy, performance, and probability tradeoffs for energy-efficient probabilistic CMOS circuits, *IEEE Trans. Circuits and Systems I*, vol. 55, no. 8, pp.2249–2262, 2008. DOI: 10.1109/TCSI.2008.920139 2

[65] R. Krieger, PLATO: A tool for computation of exact signal probability, *Proc. VLSI Design Conf.*, pp.65–68, 1993. DOI: 10.1109/ICVD.1993.669640 93

[66] D.P. Kroese, T. Taimre, and Z. I. Botev, *Handbook of Monte Carlo Methods*, New York, Willey, 2011. DOI: 10.1002/9781118014967 6, 8, 112

[67] P. J. Kuekes, W. Robinett, and R. S. Williams, Defect tolerance in resistor-logic demultiplexers for nanoelectronics, *Nanotechnology*, vol. 17, pp.2466–2474, 2006. DOI: 10.1088/0957-4484/17/10/006 70, 97

[68] S. Kullback, *Information Theory and Statistics*, Gloucester, MA, 1968. 41, 42, 112

[69] K. Kumakura, J. Motohisa, and T. Fukui, Formation and characterization of coupled quantum dots (CQDs) by selective area metalorganic vapor phase epitaxy, *J. Crystal Growth*, vol. 170, issue 1–4, pp. 700–704, 1997. DOI: 10.1016/S0022-0248(96)00641-0 25

[70] P. A. Lee and A. D. Stone, Universal conductance fluctuations in metals, *Phys. Rev. Lett.*, vol. 55, no.15, pp. 1622–1625, 1985. DOI: 10.1103/PhysRevLett.55.1622 32

[71] S. Lin and D. J. Costello, *Error Control Coding: Fundamentals and Applications*, Englewood Cliffs, NJ, Prentice-Hall, 2nd ed., 2004. 110

[72] H.-A. Loeliger, F. Lustenberger, M. Helfenstein, and F. Tarköy, Probability propagation and decoding in analog VLSI, *IEEE Trans. Inf. Theory*, vol. 47, no. 2, pp.837–843, 2001. DOI: 10.1109/18.910594 5

[73] Q. Lu, F. Gao, and S. Komarneni, Biomolecule-assisted synthesis of highly ordered snowflake-like structures of bismuth sulfide nanorods, *J. Am. Chem. Soc.*, vol. 126, no.1, pp. 54–55, 2004. DOI: 10.1021/ja0386389 25

[74] S. E. Lyshevski, Ed., Nano and Molecular Electronics, CRC Press, Taylor & Francis Group, Boca Raon, FL, 2007. 20

[75] S. E. Lyshevski, V. P. Shmerko, M. A. Lyshevski, and S. N. Yanushkevich, Neuronal processing, reconfigurable neural networks and stochastic computing, *Proc. 8th IEEE Conf. Nanotechnology*, Arlington, TX, 2008. DOI: 10.1109/NANO.2008.216 8, 13, 14, 71

[76] S. E. Lyshevski, S. N. Yanushkevich, V. P. Shmerko, and V. Geurkov, Computing paradigms for logic nanocells, *J. Computational and Theoretical Nanoscience*, vol. 5, pp.2377–2395, 2008. DOI: 10.1166/jctn.2008.1205

[77] F. J. MacWilliams and N. J. A. Sloane, *The Theory of Error-Correcting Codes*, North-Holland, Amsterdam, 1997.

[78] A. Majumdar, and S. B. K. Vrudhula, Analysis of signal probability in logic circuits using stochastic models, *IEEE Trans. VLSI*, vol. 1, no. 3, pp.365–379, 1993. DOI: 10.1109/92.238448 113

[79] D. Marculescu, R. Marculesku, and M. Pedram, Information theoretic measures for power analysis, *IEEE Trans. Computer Aided Design Integrated Circuits and Systems*, vol. 15, no. 6, pp.599–610, 1996. DOI: 10.1109/43.503930 7, 35, 41

[80] A. Marshall, *Mismatch and Noise in Modern IC Processes*, Morgan & Claypool, 2009.

[81] R. McEliece, D. MacKay, and J. Cheng, Turbo decoding as an instance of Pearl's belief propagation algorithm, *IEEE J. Selected Areas in Communication*, vol. 16, no. 2, pp.140–152, 1997. DOI: 10.1109/49.661103 17, 115

[82] C. Mead, Neuromorphic electronic systems, *Proceedings IEEE*, vol. 78, no 10, pp. 1629–1639, 1990. DOI: 10.1109/5.58356 73, 112, 115

[83] S. Minato, N. Ishiura, and S. Yajima, Shared binary decision diagram with attributed edges for efficient Boolean function manipulation, *Proc. 27th ACM/IEEE Design Automation Conf.*, 1990. DOI: 10.1145/123186.123225 59, 60, 63

[84] D. C. Montgomery and G. C. Runger, *Applied Statistics and Probability for Engineers*, 5th ed., Willey, 2011. 4, 38, 44, 49

[85] T. K. Moon, *Error Correction Coding*, Wiley, 2005. DOI: 10.1002/0471739219 87, 101

[86] E. F. Moore and C. E. Shannon, Reliable circuits using less reliable relays, *J. Franklin Institute*, vol. 262, pp. 191–208, Sept, 1956, and pp. 281–297, Oct. 1956. DOI: 10.1016/0016-0032(56)90559-2 111, 112

[87] G. E. Moore, Cramming more components onto integrated circuits, *Electronics Magazine*, vol. 38, no. 8, pp. 114-117, 1965. DOI: 10.1109/JPROC.1998.658762 20

[88] A. A. Mullin, Stochastic combinational relay switching circuits and reliability, *IRE Trans. Circuit Theory*, vol. 6, no. 1, pp.131–133, 1959. DOI: 10.1109/TCT.1959.1086521 5, 111, 112

[89] T. Muramatsu, K. Miura, Y. Shiratori, Z. Yatabe, and S. Kasai, Characterization of low-frequency noise in etched GaAs nanowire field-effect transistors having SiNx gate insulator, *Jpn. J. Appl. Phys.*, vol. 51, no.6, art. 06FE18, 2012. DOI: 10.1143/JJAP.51.06FE18 32

[90] A. F. Murray and A. V. W. Smith, Asynchronous VLSI neural networks using pulse-stream arithmetic, *IEEE J. of Solid*, vol. 23, no. 3, pp.688–697, 1988. DOI: 10.1109/4.307 5, 6

[91] P. Myllymäki, Massively parallel probabilistic reasoning with Boltzmann machines, *Applied Intelligence*, no. 11, pp. 31–44, 1999. DOI: 10.1023/A:1008324530006 8

[92] M. Nagata, T. Okumoto, and K. Taki, A Built-in Technique for Probing Power Supply and Ground Noise Distribution Within Large-Scale Digital Integrated Circuits, *IEEE J. Solid-State Circuits*, vol. 40, no. 4, pp. 813-819, 2005. DOI: 10.1109/JSSC.2005.845559 33

[93] T. Nakamura, Y. Abe, S. Kasai, H. Hasegawa, and T. Hashizume, Properties of a GaAs single electron path switching node device using a single quantum dot for hexagonal BDD quantum circuits, *J. Physics, Conf. Series*, no. 38, pp.104–107, 2006. DOI: 10.1088/1742-6596/38/1/026 21, 29

[94] T. Nakamura, Y. Shiratori, S. Kasai, and T. Hashizume, Fabrication and characterization of a GaAs-based three-terminal nanowire junction device controlled by double schottky wrap gates, *Appl. Phys. Lett.*, vol. 90, no.10 art. 102104, 2007. DOI: 10.1063/1.2711374 21

[95] S. Nakanishi, K. Fukami, T. Tada, and Y. Nakato, Metal latticeworks formed by self-organization in oscillatory electrodeposition, *J. Am. Chem. Soc.*, vol.126, no.31, pp. 9556–9557, 2004. DOI: 10.1021/ja047042y 25

[96] T. Nakanishi, E. Tsutsumi, K. Masunaga, A. Fujimoto, and K. Asakawa, Transparent aluminum nanomesh electrode fabricated by nanopatterning using self-assembled nanoparticles, *Appl. Phys. Express*, vol. 4, no.2, art. 025201, 2011. DOI: 10.1143/APEX.4.025201 23

[97] S. Narayanan, G. V. Varatkar, D. L. Jones, and N. R. Shanbhag, Computation as Estimation: A general framework for robustness and energy efficiency in SoCs, *IEEE Trans. Signal Processing*, vol. 58, no. 8, pp. 4416–4421, 2010. DOI: 10.1109/TSP.2010.2049567 17, 45, 115

[98] K. Nepal, R. I. Bahar, J. Mundy, W. R. Patterson, and A. Zaslavsky, Designing logic circuits for probabilistic computation in the presence of noise, *Proc. 42nd Design Automation Conf.*, ACM Press, pp.485–490, 2005. DOI: 10.1109/DAC.2005.193858 15, 70

[99] K. Nepal, R. I. Bahar, J. Mundy, W. R. Patterson, and A. Zaslavsky, MRF reinforcer: A probabilistic element for space redundancy in nanoscale circuits, *IEEE Micro*, Sept.-Oct., pp.19–27, 2006. DOI: 10.1109/MM.2006.96 15, 70

[100] K. Nepal, R. I. Bahar, J. Mundy, W. R. Patterson, and A. Zaslavsky, Designing nanoscale logic circuits based on Markov random fields, *J. Electronic Testing: Theory and Applications*, vol. 23, pp.255–266, 2007. DOI: 10.1007/s10836-006-0553-9 xvi, 8, 11, 13, 14, 46, 49, 52, 53, 54, 57, 58, 59, 60, 62, 63, 64, 70

[101] M. Ohtsu, K. Kobayashi, T. Lawazoe, S. Sangu, and T. Yatsui, Nanophotonics: Design, fabrication, and operation of nanometric devices using optical near fields, *IEEE J. Selected Topics in Quantum Electronics*, vol. 8, no.4, pp. 839–862, 2002. DOI: 10.1109/JSTQE.2002.801738 21

[102] K. V. Palem, Energy aware computing through probabilistic switching: a study of limits, *IEEE Trans. Comput.*, vol. 54, no. 9, pp.1123–1137, 2005. DOI: 10.1109/TC.2005.145 113

[103] K. P. Parker and E. J. McCluskey, Probabilistic treatment of general combinational networks, *IEEE Trans. Comput.*, vol. 24, no. 6, pp.668–670, 1975. DOI: 10.1109/T-C.1975.224279 92, 112

[104] J. Pearl, Evidential reasoning using stochastic simulation of causal models, *Artif. Intell.*, vol. 32, pp.245–257, 1987. DOI: 10.1016/0004-3702(87)90012-9 8, 17, 112, 114

[105] J. Pearl, *Probabilistic Reasoning in Intelligent Systems*, Morgan Kaufmann, San Mateo, CA, 1988. 35, 41, 112, 114

[106] M. A. Perkowski, M. Chrzanowska-Jeske, and Y. Xu, Lattice diagrams using Reed—Muller logic. *Proc. 3rd Int. Workshop on Applications of the Reed—Muller Expansion in Circuit Design.* Oxford, UK, pp. 85–102, 1997. 27

[107] D. K. Pradhan and S. M. Reddy. Error-control techniques for logic processors. *IEEE Trans. Comput.*, vol. 21, no. 12, pp.1331–1336, 1972. DOI: 10.1109/T-C.1972.223504 87, 103

[108] J. G. Proakis and M. Salehi, *Fundamentals of Communication Systems*, Pearson Prentice Hall, Upper Saddle River, New Jersey, 2005. 87, 98, 110

[109] Predectivie technology models, `http://ptm.asu.edu`, Arizona State University, 2008. [Online; accessed March 2012]. 105

[110] S. Ramprasad, N. R. Shanbhag, and I. N. Hajj, Information-theoretic bounds on average signal transition activity, *IEEE Trans. VLSI Systems*, vol. 7, no. 3, pp.359–368, 1999. DOI: 10.1109/92.784097 42, 115

[111] T. J. Ross, *Fuzzy Logic with Engineering Applications*, 2nd Ed., Wiley, 2004. 35

[112] A. S. Sadek, K. Nikolić, and M. Forshaw, Parallel information and compuation with restriction for noise-tolerant nanoscale logic networks, *Nanotechnology*, no. 15, pp.192–210, 2004. DOI: 10.1088/0957-4484/15/1/037 6, 42, 112

[113] M. Saitoh, N. Takahashi, H. Ishikuro and T. Hiramoto, Large electron addition energy above 250 meV in a silicon quantum dot in a single-electron transistor, *Jpn. J. Appl. Phys.*, vol. 40, no.38, pp.2010–2012, 2001. DOI: 10.1143/JJAP.40.2010 29

[114] R. Sarpeshkar, *Ultra Low Power Bioelectronics: Fundamentals, Biomedical Applications, and Bio-Inspired Systems*, Cambridge University Press, 2010. DOI: 10.1017/CBO9780511841446 2

[115] T. Sasao and J. Butler, A design method for look-up table type FPGA by pseudo-Kronecker expansion, *Proc. 23rd Int. Symp. Multiple-Valued Logic,* pp. 97–106, 1994. DOI: 10.1109/ISMVL.1994.302215 27

[116] T. Sato, A. Inoue, T. Shiota, T. Inoue, Y. Kawabe, T. Hashimoto, T. Imamura, Y. Murasaka, M. Nagata, and A. Iwata, On-die supply-voltage noise sensor with real-time sampling mode for low-power processor application, *Digest of Technical Papers of 2007 IEEE Int. Solid-State Circuit Conf. (ISSCC)* pp.290–291, 2007. DOI: 10.1109/ISSCC.2007.373408 32

[117] S. Sato, A. Kawabata, M. Nihei, and Y. Awano, Growth of diameter-controlled carbon nanotubes using monodisperse nickel nanoparticles obtained with a differential mobility analyzer, *Chem. Phys. Lett.,* vol. 382, no.3–4, pp. 361–366, 2003. DOI: 10.1016/j.cplett.2003.10.076 23

[118] T. Sawada, T. Toshikawa, K. Yoshikawa, H. Takata, K. Nii, and M. Nagata, immunity evaluation of SRAM core using DPI with on-chip diagnosis structures, *Proc. 8th Workshop Electromagnetic Compatibility of Integrated Circuits,* pp.65–70, 2011. 32

[119] A. S. Sedra and K. C. Smith, *Microelectronic Circuits,* Oxford University Press, New York, 2004. 65

[120] G. A. Shafer, *A Mathematical Theory of Evidence,* Princeton University Press, New York, 1976. 37, 112

[121] N. R. Shanbhag, Reliable and efficient system-on-chip design, *Computer,* pp.42–50, March 2004. DOI: 10.1109/MC.2004.1274003 17, 43, 45

[122] C. Shannon, Reliable machines from unreliable components, in *Notes by W. W. Peterson of Seminar at MIT,* March, 1956. 94, 111, 112

[123] A. Shen and S. Devadas, Probabilistic manipulation of Boolean functions using free Boolean diagrams. *IEEE Trans. Computer-Aided Design of Integrated Circuits and Systems,* vol. 14, no. 1, pp.87–94, 1995. DOI: 10.1109/43.363122 7, 94

[124] P. P. Shenoy, *Using Dempster-Shafer's belief-function theory in expert systems,* In: R. R. Yager, M. Fedrizzi, and J. Kacprzyk, Eds., *Advances in The Dempster-Shafer Theory of Evidence,* Willey, pp.395–439, 1994. 35

[125] V. P. Shmerko, S. N. Yanushkevich, and S. E. Lyshevski, *Computer Arithmetics for Nanoelectronics,* Taylor & Francis/CRC Press, Boca Raton, FL, 2009. xvi, 23, 28, 35, 51, 65, 67, 92, 115

[126] V. P. Shmerko, *Engineering Statistics,* Lectures for undegraduate students, 2008. 9

[127] I. Shorubalko, H. Q. Xu, I. Maximov, P. Omiling, L. Samuelson, and W. Seifert, Nonlinear operation of GaInAs/InP-based three-terminal ballistic junctions, *Appl. Phys. Lett.,* vol. 79, no.9, pp.1384–1386, 2001. DOI: 10.1063/1.1396626

[128] F. Somenzi, Cudd: Cu decision diagram package release 2.4.2. `http://vlsi.colorado.edu/~fabio/CUDD/`. 100, 101, 102

[129] M. P. Stoykovich, M. Müller, S.-O. Kim, H. H. Solak, E. W. Edwards, J. J. de Pablo, and P. F. Nealey, Directed assembly of block copolymer blends into nonregular device-oriented structures, *Science*, vol. 380, no.5727, pp.1442–1446, 2005. DOI: 10.1126/science.1111041 23

[130] K. Strehl, C. Moraga, K.-H. Temme, and R. S. Stanković, Fuzzy decision diagrams for the representation, analysis and optimization of rule bases, *Proc. 30th. Int. Sym. Multiple-valued Logic*, Portland, Oregon, pp.127–132, 2000. DOI: 10.1109/ISMVL.2000.848610 35

[131] M. Switkes, R. R. Kunz, M. Rothschild, R. F. Sinta, M. Yeung, and S.-Y. Baek, Extending optics to 50 nm and beyond with immersion lithography, *J. Vac. Sci. & Technol. B*, vol.21, no. 6, pp. 2794–2799, 2003. DOI: 10.1116/1.1624257 22

[132] M. Takeda and J. W. Goodman, Neural Networks For Computation: Number Representations and Programming Complexity, *Applied Optics*, 25, pp.3033–3046, 1986. DOI: 10.1364/AO.25.003033 84

[133] M. A. Thornton and V.S.S. Nair, Efficient calculation of spectral coefficients and their applications. *IEEE Trans. Computer-Aided Design of Integrated Circuits and Systems*, vol. 14, pp.1328–1341, 1995. DOI: 10.1109/43.469660 7, 93, 94, 113

[134] M.A. Thornton, D. M. Miller, and R. Drechsler, *Spectral Techniques in VLSI CAD*, Kluwer, Dordrecht, 2001. DOI: 10.1007/978-1-4615-1425-1 7, 94

[135] A. H. Tran, S. N. Yanushkevich, S. E. Lyshevski and V. P. Shmerko, Design of neuromorphic logic networks and fault-tolerant computing, *Proc. 11th IEEE Int. Conf. Nanotechnology*, Portland, Aug. 15–18, 2011. DOI: 10.1109/NANO.2011.6144380 8, 13, 14, 15, 46

[136] T. Tsunomura, A. Nishida, and T. Hiramoto, Verification of threshold voltage variation of scaled transistors with ultra large-scale device matrix array test element group, *Jpn. J. Appl. Phys.*, vol. 48, no.12, art. no. 124505, 2009. DOI: 10.1143/JJAP.48.124505 33

[137] D. E. Van Den Bout and T. K. Miller III, A digital architecture employing stochastism for the simulation of Hopfield neural nets, *IEEE Trans. Circuits and Syst.*, vol. 36, no. 5, pp.32–738, 1989. DOI: 10.1109/31.31321 5, 6

[138] B. J. Van Wees, L. P. Kouwenhoven, E. M. M. Willems, C. J. P. M. Harmans, J. E. Mooij, H. van Houten, C. W. J. Beenakker, J. G. Williamson, and C. T. Foxon, Quantum ballistic and adiabatic electron transport studied with quantum point contact, *Phys. Rev. B*, vol. 43, no.12, pp. 12431–12453, 1991. DOI: 10.1103/PhysRevB.43.12431 32

[139] G. V. Varatkar, S. Narayanan, N. R. Shanbhag, and D. L. Jones, Stochastic networked computation, *IEEE Trans. VLSI Systems*, vol. 18, no. 10, pp.1421–1432, 2010. DOI: 10.1109/TVLSI.2009.2024673

[140] G. V. Varatkar, S. Narayanan, N. R. Shanbhag, and D. L. Jones, Stochastic networked computation, *IEEE Trans. VLSI Systems*, vol. 18, no. 10, pp.1421–1432, 2010. DOI: 10.1109/TVLSI.2009.2024673 17

[141] P. Varshney, C. Hartmann, and J. De Faria, Application of information theory to sequential fault diagnosis, *IEEE Trans. Comput.*, vol. 31, pp.164–170, 1982. DOI: 10.1109/TC.1982.1675963 6, 7, 94, 113

[142] C. Visweswariah, Death, taxes and failing chips, *Proc. IEEE Design Automation Conf.*, Anaheim, CA, pp.343–347, 2003. DOI: 10.1109/DAC.2003.1219021

[143] J. Von Neumann, Probabilistic logics and the synthesis of reliable organisms from unreliable components, in *Automata Studies*, C. E. Shannon and J. McCarthy, Eds., Princeton University Press, Princeton, NJ, pp.43–98, 1956. 6, 88, 91, 111, 112

[144] S. B. K. Vrudhula, M. Pedram, and Y-T Lai, Edge valued binary decision diagrams, In: T. Sasao and M. Fujita, Eds, *Representations of Discrete Functions,* Kluwer, Dordrecht, pp.109–132, 1996. DOI: 10.1007/978-1-4613-1385-4 92

[145] I-C. Wey, Y-G. Chen, C.-H. Yu, et. al., Design and implementation of cost-effective probabilistic-based noise-talerant VLSI circuits, *IEEE Trans. Circuits and Systems-I*, vol. 56, no. 11, pp.2411–2424, 2009. DOI: 10.1109/TCSI.2009.2015648 8, 13, 15, 46, 49, 57, 58, 62, 63

[146] S. Winograd and J. D. Cowan, *Reliable Computation in the Presence of Noise*, MIT Press, Cambridge, MA, 1963. 91, 112

[147] X.W. Wu and F.P. Prosser, CMOS ternary logic circuits, *IEE Proceedings*, vol. 137, Pt. G, no. I, pp.20–27, 1990. 65, 66

[148] Z. Wu, Z. Chen, X. Du, J. M. Logan, J. Sippel, M. Nikolou, K. Kamaras, J. R. Reynolds, D. B. Tanner, A. F. Hebard, and A. G. Rinzler, Transparent, conductive carbon nanotube films, *Science*, vol. 305, no.5688, pp.1273–1276, 2004. DOI: 10.1126/science.1101243 23

[149] H. Q. Xu, Electrical properties of three-terminal ballistic junctions, *Appl. Phys. Lett.*, vol. 78, no.14, pp.2064–2066, 2001. DOI: 10.1063/1.1360229 21

[150] T. Yamada, M. Akazawa, T. Asai, and Y. Amemiya, Boltzmann machine neural network devices using single-electron tunnelling, *Nanotechnology*, no. 12 pp.60–67, 2001. DOI: 10.1088/0957-4484/12/1/311 13

[151] H. Yan, R. He, J. Johnson, M. Law, R. J. Saykally, and P. Yang, Dendritic nanowire ultraviolet laser array, *J. Am. Chem. Soc.*, no. 125, pp.4728–4729, 2003. DOI: 10.1021/ja034327m 25

[152] S. N. Yanushkevich, O. R. Boulanov, V. P. Shmerko, Embedding and Assembling Techniques for Spatial Computing Structure Design using Decision Trees and Diagrams, *IEEE 36th Int. Symposium Multi-Valued Logic*, Singapore, May, 2006. DOI: 10.1109/ISMVL.2006.20 116

[153] S. N. Yanushkevich and V. P. Shmerko, *Introduction to Logic Design*, Taylor & Francis Group, Boca Raton, FL, 2008. xv, 36, 56, 92, 113

[154] S. N. Yanushkevich, D. M. Miller, V. P. Shmerko, and R. S. Stanković, *Decision Diagram Techniques for Micro- and Nanoelectronic Design Handbook*, CRC Press, Taylor & Francis Group, Boca Raton, FL, 2006. xvi, 7, 23, 35, 36, 42, 53, 55, 56, 63, 92, 93, 97

[155] S. N. Yanushkevich, *Advanced Logic Desig of Micro- and Nano-electtronic Device*, Lectures for graduate students, Private communication, 2008.

[156] S. N. Yanushkevich, *Fundamentals of Biometric System Design*, Lectures for undegraduate students, Private communication, 2011.

[157] S. N. Yanushkevich, V. P. Shmerko, and S. E. Lyshevski, Three dimensional computing nanostructures, In: *Encyclopedia of Nanoscience and Nanothechnology*, H. S. Nalwa, Ed., American Scientific Publishers, vol 24, pp.445–466, 2011. 116

[158] S. N. Yanushkevich, G. Tangim, S. Kasai, S. E. Lyshevski, and V. P. Shmerko, Design of nanoelectronic ICs: Noise-tolerant logic based on cyclic BDD, *Proc. IEEE Nanotechnology Conf.*, Birmingham, UK, 2012. 46, 49, 54, 58, 62, 63, 113

[159] M. Yumoto, S. Kasai, and H. Hasegawa, Gate control characteristics in GaAs nanometer-scale Schottky wrap gate structures, *Appl. Surf. Scie.*, vol. 190, no.1, pp. 242–246, 2002. DOI: 10.1016/S0169-4332(01)00890-X 26, 28

[160] W. Zhang, K. Itoh, J. Tanida, and Y. Ichioka, Hopfield model with multistate neurons and its optoelectronic implementation. *Applied Optics*, vol. 30, no. 2, pp.195–200, 1991. DOI: 10.1364/AO.30.000195 84

[161] H.-Q. Zhao, S. Kasai, Y. Shiratori, and T. Hashizume, A binary-decision-diagram-based two-bit arithmetic logic unit on a GaAs-based regular nanowire network with hexagonal topology, *Nanotechnology*, vol.20, no. 24, art.no. 245203, 2009. DOI: 10.1088/0957-4484/20/24/245203 24, 26, 106

[162] J. Zhou, Y. Ding, S. Z. Deng, L. Gong, N. S. Xu, and Z. K. Wang, Three-dimensional tungsten oxide nanowire networks, *Advanced Materials*, vol.17, no. 17, pp.2107–2110, 2005. DOI: 10.1002/adma.200500885 25

Authors' Biographies

S. N. YANUSHKEVICH

S. N. Yanushkevich received M.Sc. (1989) and Ph.D. (1992) degrees in electrical and computer engineering from the State University of Informatics and Radioelectronics, Minsk, Belarus, and she received her Habilitation degree (1999) from the University of Technology, Warsaw, Poland. In 2001, she joined the Department of Electrical and Computer Engineering, University of Calgary, Canada, where she is currently Associate Professor. Dr. Yanushkevich is a Senior Member of the IEEE and a Member of the Institute of Electronics, Information and Communication Engineers (IEICE), Japan. In 2011, she was awarded the Invitation Fellowship by the Japan Society for Promotion of Science (JSPS) and worked at the Research Center for Integrated Quantum Electronics, Hokkaido University. Dr. Yanushkevich took part in several projects on the development of high-performance parallel computing tools for image processing applications. Her current research interests include nanocomputing in 3D and artificial intelligence in decision making. Dr. Yanushkevich serves as a Guest Editor of several international journals, and she was a general chair and co-chair of about 20 international conferences, symposia, and workshops. Dr. Yanushkevich has published over 200 technical papers and patents, and authored, co-authored, and edited books.

S. KASAI

S. Kasai received B.E., M.E., and Ph.D. degrees in electrical engineering from Hokkaido University, Hokkaido, Japan, in 1992, 1994, and 1997, respectively. He joined Optoelectronics and High Frequency Device Research Laboratories, NEC, Japan, in 1997. In 1999, he moved to the Graduate School of Electronics and Information Engineering, Hokkaido University, as a Research Associate and he has been an Associate Professor since 2001. From 2004, he has been an Associate Professor in Graduate School of Information Science and Technology and Research Center for Integrated Quantum Electronics (RCIQE). From 2007 to 2011, he has also been a researcher of PRESTO, Japan Science and Technology Agency (JST). His current research interests include III-V compound semiconductor nanodevices and their integrations, and stochastic resonance device. He is a member of IEEE, the Institute of Electronics, Information and Communication Engineers (IEICE), and the Japan Society of Applied Physics (JSAP).

G. TANGIM

G. Tangim is pursuing an M.Sc. in Electrical and Computer Engineering (ECE) in the Department of Electrical and Computer Engineering, University of Calgary. His research interest includes Fault-Tolerant Nanoarchitectures, VLSI, and Computer Aided Design of Digital logic circuits. Tangim received his B.Sc. in Electrical and Electronics Engineering from Islamic University of Technology (IUT), Bangladesh. He served as a junior lecturer in the School of Engineering and Computer Science (SECS) in Independent University, Bangladesh (IUB) for two years. He is a student member of IEEE.

T. MOHAMED

T. Mohamed is a Ph.D. candidate at the Electrical and Computer Engineering (ECE) department, University of Calgary. He obtained his M.Sc. degree from Ain Shams University, Cairo, Egypt, in the topic of reconfigurable architectures for software defined radio. He has several years of industrial experience in the implementation of algorithms for digital image processing in the bio-medical field, video surveillance, and digital video broadcast. His current research interests are focused on the design and synthesis of practical fault-tolerant circuits that can be implemented in nanotechnology. He is an IEEE member.

A.H. TRAN

A. H. Tran received M.Eng. (2007) and M.Sc. (2011) degrees in electrical and computer engineering from the Schulich School of Engineering, University of Calgary, Calgary, Alberta. His research interests include Stochastic and Probabilistic Logic Computing, Simulation of Logic Computing under extreme noise, and applying Artificial Intelligence in Fault-Tolerant computing. He is planning to pursue his Ph.D. in the near future.

V.P. SHMERKO

V. P. Shmerko is an Adjunct Professor of the Department of Electrical and Computer Engineering, University of Calgary, Canada. He is a Fellow of the Institution of Engineering and Technology (ITE), UK, and an ITE Chartered Engineer. His work in design of noise-immune and fault-tolerant discrete devices has spanned more than 30 years and has included the study of methods for modeling of computing in presence of noise caused by extreme cases of exploitation. His recent work has focused on 3D computing and probabilistic modeling of nanoscaled computer devices. Dr. Shmerko has published over 300 technical papers, research reports, authored and co-authored several books, including textbooks for undergraduate and graduated students; he was co-inventor of 45 patents on noise-tolerant computing and modeling systems.